Contents

D0495351

WITHDRAWN FROM LIBRARY

Cooperating organisations		**6**
Acknowledgements		**7**
Preface		**8**
Introduction		**9**
Chapter 1	**General requirements**	**11**
1.1	Safety	11
1.2	Required competence	12
1.3	The client	12
	1.3.1 Certificates and Reports	12
	1.3.2 Rented domestic and residential accommodation	12
1.4	Additions and alterations	13
1.5	Record keeping	13
Chapter 2	**Initial verification**	**15**
2.1	Purpose of initial verification	15
2.2	Certificates	16
2.3	Required information	17
2.4	Scope	17
2.5	Frequency of subsequent inspections	17
2.6	Initial inspection	18
	2.6.1 General procedure	18
	2.6.2 Comments on individual items to be inspected	18
	2.6.3 Inspection checklist	25
2.7	Initial testing	31
	2.7.1 Test results	31
	2.7.2 Electrical Installation Certificate	31
	2.7.3 Model forms	32
	2.7.4 The sequence of tests	32
	2.7.5 Continuity of protective conductors, including main and supplementary bonding	32
	2.7.6 Continuity of ring final circuit conductors	35
	2.7.7 Insulation resistance	39
	2.7.8 Confirming SELV or PELV circuits by insulation testing	42
	2.7.9 Testing of electrically separated circuits	43

	2.7.10	Protection by barriers or enclosures provided during erection	44
	2.7.11	Proving and testing of non-conducting location (insulation resistance/impedance of floors and walls)	45
	2.7.12	Polarity testing	47
	2.7.13	Earth electrode resistance testing	48
	2.7.14	Protection by automatic disconnection of supply	51
	2.7.15	Earth fault loop impedance verification	52
	2.7.16	Prospective fault current, I_{pf}	54
	2.7.17	Phase sequence testing	57
	2.7.18	Operational and functional testing of RCDs	58
	2.7.19	Other functional testing	59
	2.7.20	Verification of voltage drop	59
	2.7.21	Verification in medical locations	60
	2.7.22	Verification of electromagnetic disturbances	60

Chapter 3 Periodic inspection and testing — 61

3.1	Purpose of periodic inspection and testing	61
3.2	Necessity for periodic inspection and testing	61
3.3	Electricity at Work Regulations	62
3.4	Design	62
3.5	Routine checks	63
3.6	Required information	63
3.7	Frequency of periodic inspections	64
3.8	Requirements for inspection and testing	66
	3.8.1 Scope	66
	3.8.2 Process – prior to carrying out inspection and testing	66
	3.8.3 General procedure	67
	3.8.4 Setting inspection and testing samples	68
3.9	Periodic inspection	70
	3.9.1 Example checklist of items that require inspection	70
3.10	Periodic testing	75
	3.10.1 General	75
	3.10.2 Tests to be made	75
	3.10.3 Additional notes on periodic testing	76
3.11	Electrical Installation Condition Report	79
3.12	Periodic inspection of installations to an earlier edition of BS 7671 or the IEE Wiring Regulations	80

Chapter 4 Test instruments and equipment — 81

4.1	Instrument standard	81
4.2	Instrument accuracy	81
4.3	Low-resistance ohmmeters	82
4.4	Insulation resistance testers	83
4.5	Earth fault loop impedance testers	83
4.6	Earth electrode resistance testers	84

4.7	RCD testers	84
4.8	Phase rotation instruments	84
4.9	Thermographic equipment	85

Chapter 5 Forms   **89**
5.1	Initial verification (inspection and testing) forms	89
5.2	Minor works	89
5.3	Periodic inspection and testing	90
5.4	Model forms for certification and reporting	90

Appendix A Maximum permissible measured earth fault loop impedance   **111**
A1	Tables	111
A2	Appendix 14 of BS 7671:2008	114
A3	Methods of adjusting tabulated values of Z_s	115

Appendix B Resistance of copper and aluminium conductors   **119**
| B1 | Standard overcurrent devices | 121 |
| B2 | Steel-wire armour, steel conduit and steel trunking | 122 |

Index   **123**

Cooperating organisations

The Institution of Engineering and Technology acknowledges the invaluable contribution made by the following organisations in the preparation of this Guidance Note.

BEAMA Installation Ltd
A. Murray MIET
P. Sayer IEng MIET GCGI

City and Guilds
M. Doughton IEng MIET LCG

EAL
K. Sparrow

Electrical Contractors' Association
G. Digilio IEng FIET ACIBSE MSLL

Electrical Contractors' Association of Scotland (SELECT)
B. Cairney IEng MIET
D. Millar IEng MIET

Electrical Safety Council

Health and Safety Executive
K. Morton BSc CEng FIET

Highway Electrical Association
G. Pritchard BTech(Hons) CEng MIET
    FILP TechIOSH

IET – members of the Users Forum
S. Brand
A. Jewsbury
J. Peckham MIET LCGI
O. Smith BSc(Hons) MIET MSLL

Institution of Engineering and Technology
P.R.L. Cook CEng FIET
P.E. Donnachie BSc CEng FIET
H.R. Lovegrove IEng FIET
Eur Ing L. Markwell MSc BSc(Hons)
    CEng MIEE MCIBSE LCGI
Eur Ing J. Pettifer BSc CEng MIET FCQI
I.M. Reeve BTech CEng MIEE
R. Townsend MIET

National Association for Professional Inspectors and Testers
B. Allan BEng(Hons) CEng MIET
F. Bertie MIET

NICEIC
J.M. Maltby-Smith BSc(Hons) PG Dip
    Cert Ed IEng MIET

IET – editor and contributor
Eur Ing D. Locke BEng(Hons) CEng
    MIEE MIEEE

Acknowledgements

References to British Standards, CENELEC Harmonization Documents and International Electrotechnical Commission (IEC) standards are made with the kind permission of BSI. Complete copies can be obtained by post from:

BSI Customer Services
389 Chiswick High Road
London W4 4AL
Tel: +44 (0)20 8996 9001
Fax: +44 (0)20 8996 7001
Email: orders@bsi-global.com

The BSI also maintains stocks of international and foreign standards, with many English translations. Up-to-date information on BSI standards can be obtained from the BSI website: www.bsi-global.com

Illustrations of test instruments were provided by Farquhar Design: www.farquhardesign.co.uk

Thermographic surveying images are reproduced with the kind permission of FLIR Systems Ltd: www.flirthermography.co.uk

Cover design and illustration were created by The Page Design: www.thepagedesign.co.uk

It is strongly recommended that anyone involved in work on or near electrical installations possesses a copy of the *Memorandum of guidance on the Electricity at Work Regulations 1989* (HSR25) published by the Health and Safety Executive.

Copies of Health and Safety Executive documents and approved codes of practice (ACOP) can be obtained from:

HSE Books
PO Box 1999
Sudbury, Suffolk CO10 2WA
Tel: +44 (0)1787 881165
Email: hsebooks@prolog.uk.com
Web: http://books.hse.gov.uk

Preface

This Guidance Note is part of a series issued by the Institution of Engineering and Technology to explain and enlarge upon the requirements in BS 7671:2008, the 17th Edition of the IET Wiring Regulations, incorporating Amendment No. 1:2011

Note that this Guidance Note does not ensure compliance with BS 7671. It is intended to explain some of the requirements of BS 7671 but readers should always consult BS 7671 to satisfy themselves of compliance.

The scope generally follows that of BS 7671; the relevant Regulations and Appendices are noted in the margin. Some Guidance Notes also contain material not included in BS 7671:2008 but which was included in earlier editions of the Wiring Regulations. All of the Guidance Notes contain references to other relevant sources of information.

Electrical installations in the United Kingdom that comply with BS 7671 are likely to satisfy the relevant parts of Statutory Regulations such as the Electricity at Work Regulations 1989, however this cannot be guaranteed. It is stressed that it is essential to establish which Statutory and other Regulations apply and to install accordingly. For example, an installation in premises subject to licensing may have requirements different from, or additional to, BS 7671 and these will take precedence.

Introduction

This Guidance Note is concerned principally with Part 6 of BS 7671 - Inspection and testing.

Neither BS 7671 nor the Guidance Notes are design guides. It is essential to prepare a full specification prior to commencement or alteration of an electrical installation.

514.9 The specification should set out the detailed design and provide sufficient information to enable competent persons to carry out the installation and to commission it. The specification must include a description of how the system is to operate and all the design and operational parameters. It must provide for all the commissioning procedures that will be required and for the provision of adequate information to the user. This should be by means of an operational manual or schedule, and 'as fitted' drawings if necessary.

It must be noted that it is a matter of contract as to which person or organisation is responsible for, in turn, the design, specification, construction and verification of the installation and the provision of any operational information.

The persons or organisations who may be concerned in the preparation of the specification include:

The Designer
The Installer
The Verifier
The Distributor of Electricity
The Installation Owner and/or User
The Architect
Area Building Control bodies
Any Regulatory Authority
Any Licensing Authority
The Health and Safety Executive.

133.1
Sect 511 Precise details of each item of equipment should be obtained from the manufacturer and/or supplier and compliance with appropriate standards confirmed.

The operational manual must include a description of how the system as installed is to operate and all commissioning records. The manual should also include manufacturers' technical data for all items of switchgear, luminaires, accessories, etc. and any special instructions that may be needed. The Health and Safety at Work etc. Act 1974 Section 6 and the Construction (Design and Management) Regulations 2007 are concerned with the provision of information, and guidance on the preparation of technical manuals is given in the BS 4884 series *Technical manuals* and BS 4940 series *Technical information on construction products and services*. The size and complexity of the installation will dictate the nature and extent of the manual.

© The Institution of Engineering and Technology

General requirements

1

1.1 Safety

Electrical testing inherently involves some degree of hazard. It is the inspector's duty to ensure his (or her) own safety, and that of others, in the performance of the test procedures. The safety procedures detailed in Health and Safety Executive Guidance Note GS38 (revised) 'Electrical test equipment for use by electricians' should be observed. Where testing does not require the equipment or part of an installation to be live, it should be made dead and safely isolated. Guidance on live and dead working can be found in the Health and Safety Executive publication *Memorandum of guidance on the Electricity at Work Regulations 1989* (HSR 25). Guidance on safe isolation procedures can be found in a best practice guide published by the Electrical Safety Council.

When using test instruments, safety is best achieved by precautions such as:

1 understanding the equipment to be used, its rating and the characteristics of the installation it is proposed to use the equipment upon

2 checking that the instruments to be used conform to the appropriate British Standard safety specifications. These are BS EN 61010 *Safety requirements for electrical equipment for measurement, control, and laboratory use* and BS 5458:1977 (1993) *Specification for safety requirements for indicating and recording electrical measuring instruments and their accessories.* BS 5458 has been withdrawn, but is the standard to which older instruments should have been manufactured

3 checking that test leads including any probes or clips used are in good order, are clean and have no cracked or broken insulation. Where appropriate, the requirements of the Health and Safety Executive Guidance Note GS 38 should be observed for test leads. This recommends the use of fused test leads aimed primarily at reducing the risks associated with arcing under fault conditions.

Particular attention should be paid to the safety aspects associated with any tests performed with instruments capable of generating a test voltage greater than 50 V, or which use the supply voltage for the purposes of the test in earth loop testing and residual current device (RCD) testing. Note the warnings given in Section 2.7.15 and 2.7.18 of this Guidance Note.

Electric shock hazards can arise from, for example, capacitive loads such as cables charged in the process of an insulation test, or voltages on the earthed metalwork whilst conducting a loop test or RCD test. The test limits quoted in these guidelines are intended to minimise the chances of receiving an electric shock during tests.

1.2 Required competence

610.5 The inspector carrying out the inspection and testing of any electrical installation must, as appropriate to his or her function, have a sound knowledge and experience relevant to the nature of the installation being inspected and tested, and of BS 7671 and other relevant technical standards. The inspector must also be fully versed in the inspection and testing procedures and employ suitable test equipment during the inspection and testing process.

The inspector should have sufficient inspection and testing skills together with experience in interpreting the results with respect to the requirements of BS 7671.

It is worth noting that the person responsible for inspection and testing may be required to formally demonstrate competence by means of registration/certification under a recognised scheme, for example as a condition of contract and/or as a requirement of Area Building Control Bodies.

It is the responsibility of the inspector, as appropriate to either the initial or periodic inspection, to:

610.1   **1**   ensure no danger occurs to any person or livestock and property is not damaged
612.1   **2**   compare the inspection and testing results with the design criteria (where available), with BS 7671 and/or previous records, as appropriate
        3   confirm compliance or non-compliance with BS 7671
        4   take a view and report on the condition of the installation.

634.2 In the event of a dangerous situation being found, the inspector should recommend the immediate isolation of the defective part. The person ordering the work should be informed, in writing, of this recommendation without delay.

1.3 The client

1.3.1 Certificates and Reports

631.1 Following the initial verification of a new installation or changes to an existing installation,
632.1 an Electrical Installation Certificate, together with a schedule of inspections and a schedule of test results, is required to be given to the person ordering the work. In this context, 'work' means the installation work, not the work of carrying out the inspection
631.2 and testing. Likewise, following the periodic inspection and testing of an existing
634.1 installation, an Electrical Installation Condition Report, together with schedules of inspection and schedules of test results, are required to be given to the person ordering the inspection.

Sometimes the person ordering the work is not the user of the installation. In such cases it is necessary for the user (e.g. employer or householder) to have a copy of the inspection and test documentation. It is recommended that those providing documentation to the person ordering the work recommend that a copy of the forms be passed to the user, including any purchaser of a domestic property.

1.3.2 Rented domestic and residential accommodation

In England and Wales, the Landlord and Tenant Act 1985, Section 11 Repairing obligations in short leases sub-section (1)(b) implies that a landlord 'shall keep in repair and proper working order the installations in the dwelling-house for the supply of water, gas and electricity'.

A similar requirement can be found in the Housing (Scotland) Act 2006, Section 13, which states that the electrical installation must be in a reasonable state of repair in order to comply with the Act. The legislation goes on to state that the installation must be maintained.

The above Acts do not directly specify periodic inspection and testing of the electrical installation. Periodic inspection and testing is a means of demonstrating compliance with the Acts.

Any repairs must be carried out by a competent person. The landlord is responsible for confirming the competency of any contractors carrying out such work.

There are requirements for houses of multiple occupancy (HMO) and these are required to be inspected and tested at least every five years, see Chapter 3.

1.4   Additions and alterations

132.16
633.1
633.2
Every addition or alteration to an existing installation must comply with the current edition of BS 7671 and must not impair the safety of the existing installation. In this respect, in order to verify that the addition or alteration to an electrical installation complies with BS 7671, the existing installation must be inspected and tested to confirm the safety of the addition or alteration, including for example:

▶ protective conductor continuity
▶ earth fault loop impedance.

633.2
Whilst there is no obligation to inspect and test any part of the existing installation that does not affect and is not affected by the addition or alteration, observed departures are required to be noted in the comments box of electrical installation certificates (single-signature or multiple-signature) and minor works certificates.

1.5   Record keeping

132.13
514.9
It is a requirement that the appropriate documentation called for in Regulation 514.9, Part 6 and Part 7 is provided for every electrical installation.

Sect 622
Appx 6 –
guidance to
recipients
Records of all checks, inspections and tests, including test results, should be kept throughout the working life of an electrical installation. This will enable deterioration to be identified. They can also be used as a management tool to ensure that maintenance checks are being carried out and to assess their effectiveness.

For non-domestic installations, Regulation 20 (2) sub-section (e) of the Construction (Design and Management) Regulations 2007 requires a record known as 'the health and safety file' to be prepared, reviewed and updated. This should contain any information relating to the project which is likely to be needed during any subsequent construction work to ensure the health and safety of persons. Sub-section (f) requires that the health and safety file is passed on to the client on completion of the construction work.

For domestic installations the NHBC (National House-Building Council) guidance recommends that all instructions for services be passed to the building owner.

In both domestic and non-domestic cases there may be insurance requirements that imply or specify records.

Regulation 17 (3) (3) requires that reasonable steps be taken to ensure that once the construction phase has been completed the information in the health and safety file remains available for inspection by any person who might need it to comply with any relevant statutory provisions. It also requires that the file is revised and updated as often as may be appropriate to incorporate any relevant new information.

Electrical installation certificates, minor electrical works certificates and electrical installation condition reports would constitute relevant information in relation to this requirement.

The Electromagnetic Compatibility (EMC) Regulations 2006 are statutory and require that the client keep the information relating to compliance with EMC criteria for the life of the installation.

Initial verification   2

2.1   Purpose of initial verification

610.1
611.2
612.1

Initial verification is carried out on a new installation before it is put into service. The purpose of initial verification is to confirm by way of inspection and testing, during construction and on completion, that the installation complies with the design and construction aspects of BS 7671, in so far as is reasonably practicable.

Appx 6,
Intro (v)

It is important to recognise the responsibilities of the signatories for the design, construction and verification. While the inspector is responsible for verifying aspects of both design and construction, he/she cannot and is not meant to absolve responsibility for these elements; indeed, this is why the inspector carries out the inspection and testing *so far as is reasonably practicable.*

Example

Consider one aspect of the design: the inspector should check that the cable sizes, as specified, have been correctly selected and installed. In order to do this he or she would need the design criteria, say a cable schedule, and will then carry out a visual inspection of the installed cable sizes for comparison. The most logical position to do so would be at the distribution board housing the cables' protective devices. It would be unreasonable for him or her to carry out design cable sizing checks, as this is the responsibility of the designer.

At this point, it would also be unreasonable for the inspector to check that each cable size at the distribution point is maintained throughout the cable's length (this is the responsibility of the installer or constructor).

This example illustrates the principle and responsibilities, that the designer and constructor of the installation are both confirming their facets and that the inspector carries out checks but only in so far as to supplement the responsibilities of others.

611.2

BS 7671 provides a format list in Regulation 611.2 of items to be verified, again so far as reasonably practicable; these are as follows:

1   Installed electrical equipment is of the correct type and complies with applicable British Standards or acceptable equivalents
2   The fixed installation is correctly selected and erected
3   The fixed installation is not visibly damaged or otherwise defective.

Sect 611   **Inspections**

Inspections are an important element of inspection and testing, and are described in section 2.6 of this Guidance Note.

Sect 612 **Tests**

The tests are described in section 2.7 of this Guidance Note.

631.1 **Results**

The results of inspection and tests are to be recorded as appropriate. The *Memorandum of guidance on the Electricity at Work Regulations 1989* (HSR25) recommends records of all maintenance including test results be kept throughout the life of an installation – see guidance on EWR Regulation 4(2). This can enable the condition of equipment and the effectiveness of maintenance to be monitored.

612.1 **Relevant criteria**

The relevant criteria are, for the most part, the requirements of the Regulations for the particular inspection or test and most criteria are given in Chapters 2 and 3 of this Guidance Note.

There will be some instances where the designer has specified requirements which are particular to the installation concerned. For example, the intended impedances may be different from those in BS 7671. In this case, the inspector should either ask for the design criteria or forward the test results to the designer for verification with the intended design. In the absence of such data the inspector should apply the requirements set out in BS 7671.

Verification

The responsibility for comparing inspection and test results with relevant criteria, as required by Regulation 612.1, lies with the party responsible for inspecting and testing the installation. This party, which may be the person carrying out the inspection and testing, should sign the inspection and testing box of the Electrical Installation Certificate or the declaration box of the Minor Electrical Installation Works Certificate. If the person carrying out the inspection and testing has also been responsible for the design and construction of the installation, he or she must also sign the design and construction boxes of the Electrical Installation Certificate, or make use of the single-signature Electrical Installation Certificate.

2.2 Certificates

Appx 6 Appendix 6 of BS 7671 contains three model forms for the initial certification of a new installation or for an addition or alteration to an existing installation, as follows:

▶ multiple-signature Electrical Installation Certificate
▶ single-signature Electrical Installation Certificate
▶ Minor Electrical Installation Works Certificate.

Examples of typical forms are given in Chapter 5.

Multiple-signature Electrical Installation Certificate

The multiple-signature certificate allows different persons to sign for design, construction, inspection and testing, and allows two signatories for design where there is mutual responsibility. Where designers are responsible for identifiably separate parts of an installation, the use of separate forms would be appropriate.

Single-signature Electrical Installation Certificate

Where design, construction, inspection and testing are the responsibility of one person, a certificate with a single signature may replace the multiple signature form.

Minor Electrical Installation Works Certificate

This certificate is to be used only for minor works that do not include the provision of a new circuit, such as an additional socket-outlet or lighting point to an existing circuit.

The certificate may also be used for the replacement of equipment such as accessories or luminaires, but not for the replacement of distribution boards, consumer units or similar items.

2.3 Required information

610.2 BS 7671 requires that the following information shall be made available to the person or persons carrying out the inspection and testing:

Assessment of general characteristics

311.1 **1** the maximum demand, expressed in amperes, kW or kVA per phase (after diversity is taken into account)

312.1 **2** the number and type of live conductors of the source(s) of energy and of the circuits used in the installation

312.2 **3** the type of earthing arrangement used by the installation and any facilities provided by the distributor for the user

313.1 **4** the nominal voltage(s) and its characteristics including harmonic distortion

5 the nature of the current and supply frequency

6 the prospective short-circuit current at the origin of the installation

7 the earth fault loop impedance (Z_e) of that part of the system external to the installation

8 the type and rating of the overcurrent protective device acting at the origin of the installation.

Note: These characteristics should be available for all sources of supply.

Diagrams, charts or tables

The Health and Safety at Work etc Act 1974 generally requires relevant information to be available as an aid to safe use, inspection, testing and maintenance. This may

514.9.1 include those items listed in Regulation 514.9.1 as follows:

9 the type and composition of circuits, including points of utilisation, number and size of conductors and type of cable. This should include the installation method shown in Appendix 4 section 7 'Methods of installation' of BS 7671

410.3.2 **10** the method used for compliance with the requirements for basic and fault protection and, where appropriate, the conditions required for automatic disconnection

11 the information necessary for the identification of each device performing the functions of protection, isolation and switching, and its location

12 any circuit or equipment vulnerable to a particular test.

2.4 Scope

It is essential that the inspector and the person ordering the inspection agree the extent of the installation to be inspected beforehand, and any criteria regarding the limit of the inspection. The details should be recorded on the Certificate.

2.5 Frequency of subsequent inspections

The time intervals between the recommended dates of periodic inspections need
134.2.2 consideration. The date for the first periodic inspection and test is required to be considered and recommended by the installation designer, as part of his or her design.

The date of each subsequent periodic inspection is required to be considered and recommended as part of carrying out a periodic inspection and test, by the person undertaking that particular inspection and test.

2.6 Initial inspection

2.6.1 General procedure

Inspection and, where appropriate, testing should be carried out and recorded on suitable schedules progressively throughout the different stages of erection and before the installation is certified and put into service.

610.1 It should be noted that Regulation 610.1 requires inspection and testing to be carried out during the erection stage of the installation, as well as on completion.

A model Schedule of Inspections is shown in Chapter 5.

2.6.2 Comments on individual items to be inspected

611.3 BS 7671 provides a list of items considered as a minimum to be inspected as follows:

Sect 526 **a Connection of conductors**

Every connection between conductors and equipment/other conductors should provide durable electrical continuity and adequate mechanical strength.

b Identification of cables and conductors

Sect 514 It should be checked that each core or bare conductor is identified as necessary. The single colour green must not be used. The colour combination green-and-yellow is only to be used for protective conductors and single-core green-and-yellow identified conductors (other than PEN conductors) cannot be overmarked with another colour or alphanumeric symbols at terminations.

c Routing of cables

522.8 Cables and their cable management systems should be designed and installed taking into account the mechanical stresses that users of the installation will make upon the installation.

A key requirement to note is for cables installed in a wall or partition at a depth of less than 50 mm from the surface *where the installation is not intended to be under the*
522.6.101 *supervision of a skilled or instructed person*. If the cable used does not incorporate an earthed metallic covering; or, is not installed in an earthed conduit, trunking or duct; or, is not provided with mechanical protection sufficient to prevent damage being
522.6.102 caused by nails, screws or similar, or is not supplied via SELV or PELV, it will be necessary to provide additional protection by means of an RCD having a rated residual operating current not exceeding 30 mA. This is a requirement even when cables are run within the permitted cable routes described in Regulation 522.6.101 (v), as stated in Regulation 522.6.102.

522.6.103 Another requirement relates to cables installed in a wall or partition, the construction of which contains metallic component parts such as studs, frames or skins, once again *where the installation is not intended to be under the supervision of a skilled or instructed person*. Irrespective of the depth at which the cable has been installed, either the cable has an earthed metallic covering, is run in an earthed wiring system, or given adequate protection from penetration by nails, screws or similar, or additional protection by means of an RCD as described in the paragraph above will be required.

132.7
Sect 523
Sect 524
Sect 525

d Cable selection

Where practicable, the cable size should be assessed against the protective arrangement based upon information provided by the installation designer (where available).

Reference should be made, as appropriate, to Appendix 4 of BS 7671.

132.14.1
530.3.2

e Verification of polarity – single-pole device in a TN or TT system

It must be verified that single-pole devices for protection or switching are installed in line conductors only.

f Accessories and equipment

Correct connection (suitability, polarity, environmental, etc.) must be checked.

553.1.3 Table 55.1 of BS 7671 is a schedule of types of plug and socket-outlet, ratings, and the associated British Standards.

553.2.2 Particular attention should be paid to the requirements for cable couplers to ensure that female couplers are fitted at the end remote from the supply.

559.6.1.7 Bayonet lampholders B15 and B22 should comply with BS EN 61184 and be of temperature rating T2.

Sect 527

g Selection and erection to minimize the spread of fire

Fire barriers, suitable seals and/or protection against thermal effects should be provided if necessary to meet the requirements of BS 7671. These are good examples of items which can and should be inspected during the erection stage.

Each sealing arrangement should be inspected to verify that it conforms to the manufacturer's erection instructions. This may be impossible without dismantling the system and it is essential, therefore, that inspection should be carried out at the appropriate stage of the work, and that this is recorded at the time for incorporation in the inspection and test documents.

Chap 41

h Methods of protection against electric shock (protective measures)

The most common protective measure is basic protection by insulation and enclosures together with fault protection by automatic disconnection of supply. More unusual systems are discussed later.

Sect 416
416.2

(i) Basic protection

Basic protection is most usually provided by insulation and/or enclosures. The inspection of this measure is to check that insulation has not been damaged during installation and that enclosures have been selected for at least IPXXB or IP2X and, for top surfaces, at least IPXXD or IP4X, and are not damaged. (Insulation resistance is of course a fundamental test to be carried out – section 2.7.)

Note: The rarer methods of protection by obstacles or placing out of reach are discussed later in this section (see *Other measures of basic protection*).

(ii) Fault protection

The various methods of fault protection are classified in a number of subsections in BS 7671, and are:

Sect 411 **1** automatic disconnection of supply
418.1 **2** non-conducting location

418.2   **3**   earth-free local equipotential bonding

413.3   **4**   electrical separation.

Method **1** – automatic disconnection of supply (ADS)

For each circuit, automatic disconnection is required and, although the main part of verification is measurement of earth fault loop impedance in order to confirm disconnection times, there are inspection aspects to consider for verifying ADS as follows:

Presence of appropriate protective conductors:

542.3   ▶   earthing conductor

Sect 543   ▶   circuit protective conductors

Sect 544   ▶   protective bonding conductors
        –   main bonding conductors
        –   supplementary bonding conductors (where required).

312.2   The earthing system must be determined, e.g.

    ▶   TN-C-S system (protective multiple earthing (PME))

    ▶   TN-S system

    ▶   TT system (earth electrode(s)).

411.4.5   The earth fault loop impedance must be appropriate for the protective device, i.e. RCD or overcurrent device.

Sect 418   Fault protection by method **2** (non-conducting location) and method **3** (earth-free local equipotential bonding) are more specialised protection methods. It is essential that inspection of such systems be carried out by persons competent in the discipline and having adequate information on the design of the system. For these specialised systems, the designer and client will advise of, and agree, the necessary effective and continuing supervision. This will also be the case where the protective measure of electrical separation is used to supply more than one item of current-using equipment.

Method **4** (electrical separation) is discussed later in this section.

Other protective measures, providing both basic and fault protection, include SELV, PELV, double insulation and reinforced insulation.

Sect 414   For SELV and PELV, requirements include:

  1   the nominal voltage must not exceed 50 V a.c. or 120 V d.c.

  2   an isolated source, e.g. a safety isolating transformer to BS EN 61558-2-6

  3   protective separation from all non SELV or PELV circuits

  4   for SELV, basic insulation between the SELV system and earth

  5   SELV exposed-conductive-parts must have no connection with earth, exposed-conductive-parts or protective conductors of other systems.

412.1.1   For double insulation, basic protection is provided by basic insulation, and fault protection is provided by supplementary insulation.

For reinforced insulation, both basic and fault protection are provided by reinforced insulation between live parts and accessible parts.

412.1.3 Where double or reinforced insulation is to be employed as the sole protective measure, it is important to confirm that the installation or circuit so protected will remain under effective supervision to prevent any unauthorised change being made which could impair the effectiveness of the safety measure.

Other measures of basic protection

417.2 *Obstacles*

Protection by obstacles provides protection against unintentional contact and, where used, the area shall be accessible only to skilled persons or to instructed persons under their supervision. This method of protection is not to be used in some installations and locations of increased shock risk. See Part 7 of BS 7671.

417.3 *Placing out of reach*

Placing out of reach also provides basic protection. Increased distances are necessary where long or bulky conducting objects are likely to be handled in the vicinity.

410.3.5 Bare live parts are only permitted in areas accessible only to skilled persons or to instructed persons under their supervision. The dimensions of passageways should be checked against the guidance in Appendix 3 of the *Memorandum of guidance on the Electricity at Work Regulations 1989* (HSR25) issued by the Health and Safety Executive.

Sect 729 Part 729 of Amendment No 1 to BS 7671 was included as a Part 7 for special installations where open switchgear or busbars are permitted and where the area is accessible only to skilled or instructed persons.

Inspection for verification of these areas requires careful checking, including the measurement of separation distances, for example the 'arm's reach' as per Figure 417 of BS 7671; this must be confirmed with the installation isolated.

Specialised systems

410.3.6 The specialised protective measures in Section 418 and discussed below may only be employed in installations or parts thereof which remain under the supervision of skilled persons at all material times, to prevent unauthorised changes being made which may render such protective methods ineffective.

418.1 *Non-conducting location*

Where protection by this method is employed (e.g. in electronic equipment test areas)
418.1.1 all installed electrical equipment should meet the requirements of Section 416 with regard to provisions for basic protection.

418.1.2 Further, the exposed-conductive-parts of the installation should be so arranged that it is not possible for persons to make simultaneous contact with either two exposed-conductive-parts, or an exposed-conductive-part and any extraneous-conductive-part under normal operating conditions, if these parts are liable to be at different potentials as a result of failure of the basic insulation of a live part. The inspector should confirm
418.1.3 the achievement of this and check that within the location there are no protective conductors (see also the specific test for this method in 2.7.11).

418.2 *Earth-free local equipotential bonding*

The use of earth-free equipotential bonding is intended to prevent the appearance of dangerous touch voltages under fault conditions.

CITY OF LIVERPOOL COLLEGE
VAUXHALL ROAD
LIVERPOOL
L3 6BN

418.2.1 Where protection by this method is employed, all installed electrical equipment should meet the requirements of Section 416 with regard to provisions for basic protection and this can be confirmed by inspection.

418.2.2 All simultaneously accessible exposed-conductive-parts and extraneous-conductive-parts should be interconnected by equipotential bonding conductors.

418.2.3 Measures must be taken to ensure that the local equipotential bonding conductors are not connected to Earth either directly or unintentionally via the exposed- and extraneous-conductive-parts to which they are connected.

418.2.5 A warning notice complying with Regulation 514.13.2 must be fixed in a prominent position adjacent to every point of access to the location concerned. This method is sometimes combined with 'electrical separation'.

The inspection, supplemented with tests, should verify that no item is earthed within the area and that no earthed services or conductors enter or traverse the area, including the floor and ceiling. Inspection should confirm whether or not this has been achieved.

Sect 413 *Electrical separation*

418.3 This method may be applied to the supply of an individual item of equipment, or for more than one item of equipment.

413.1.1 Electrical separation is a protective measure where basic protection is provided by insulation of live parts, or by barriers/enclosures in accordance with Section 416, and fault protection is provided by simple separation of the separated circuit from other circuits and from earth.

If it is intended to supply more than one item of equipment using electrical separation, it will be necessary to meet the requirements of Regulation 418.3.

418.3.3 The separated circuit should be protected from damage and insulation failure.

418.3.4 Any exposed-conductive-parts should be connected together by insulated, non-earthed equipotential bonding conductors which should not be connected to the protective conductor or exposed-conductive-parts of any other circuit or to any extraneous-conductive-parts.

418.3.5 Socket-outlets should have a protective conductor contact, which is connected to the protective bonding system described above.

418.3.6 All flexible cables should contain a protective conductor for use as an equipotential bonding conductor, except where such a cable supplies only items of equipment having double or reinforced insulation.

418.3.7 If two faults affecting two exposed-conductive-parts occur and where conductors of different polarity feed these, a protective device should disconnect the supply in accordance with the disconnection time given in Table 41.1.

418.3.8 The product of the nominal voltage (volts) and length (metres) of the wiring system should not exceed 100,000 Vm and the length of the wiring system should not exceed 500 m.

Sect 415 **Additional protection**

415.1.1 It should be confirmed that an RCD selected to provide additional protection has a rated residual operating current ($I_{\Delta n}$) not exceeding 30 mA.

415.1.2 It should be confirmed that appropriate measures for basic and fault protection are in accordance with Sections 411 to 414, as an RCD must not be used as the sole means of protection.

415.2.1 Where supplementary bonding is provided it should encompass all simultaneously accessible exposed-conductive-parts, extraneous-conductive-parts and the protective conductors of all equipment in the location where this protective measure is being applied.

415.2.2 The effectiveness of supplementary equipotential bonding as provided may be verified where the resistance between simultaneously accessible exposed- and extraneous-conductive-parts fulfils the following condition:

$$R \leq 50 \text{ V}/I_a \text{ for a.c. systems}$$

$$R \leq 120 \text{ V}/I_a \text{ for d.c. systems}$$

where I_a is the operating current of the protective device in amps; for overcurrent devices, this is the 5 second operating current, and for RCDs, $I_{\Delta n}$.

Sect 515 **i Prevention of mutual detrimental influence**

Regulations 132.11 and 515.1 require that the electrical installation and its equipment shall not cause detriment to other electrical and non-electrical installations. The inspector is advised to step back and think about other systems whilst carrying out the inspection.

Sect 537 **j Isolating and switching devices**

BS EN 60947-1 *Specification for low voltage switchgear and controlgear – General rules* defines standard utilisation categories which allow for conditions of service use and the switching duty to be expected.

All switch utilisation categories must be appropriate for the nature of the load – see Table 2.1. It would be part of the design to specify the appropriate type of device.

GN2 Guidance Note 2: *Isolation & Switching* provides more comprehensive guidance on this subject and should be consulted and its contents taken into account.

▼ **Table 2.1** Examples of utilisation categories for alternating current installations

Utilisation category		Typical applications
Frequent operation	Infrequent operation	
AC-20A	AC-20B	Connecting and disconnecting under no-load conditions
AC-21A	AC-21B	Switching of resistive loads including moderate overloads
AC-22A	AC-22B	Switching of mixed resistive and inductive loads, including moderate overloads
AC-23A	AC-23B	Switching of motor loads or other highly inductive loads

If switchgear to BS EN 60947-1 is suitable for isolation it will be marked with the symbol:

This may be endorsed with a symbol advising of function, e.g. for a switch disconnector:

Table 53.4 Guidance on the suitability or otherwise of protective, isolation and switching devices to be employed for one or more of the functions of isolation, emergency switching and functional switching is given in Table 53.4 in BS 7671 and Guidance Note 2.

An isolation exercise should be carried out to check that effective isolation can be achieved. This should include, where appropriate, locking-off and inspection or testing to verify that the circuit is dead and no other source of supply is present.

k Presence of undervoltage protective devices

Sect 445 Suitable precautions should be in place where a reduction in voltage, or loss and subsequent restoration of voltage, could cause danger. Normally such a requirement concerns only motor circuits. If precautions are required they will have been specified by the designer; however, the devices used must be confirmed as matching the equipment specification and the relevant regulations in Section 445.

l Protective devices

Chap 43 Some protective devices have user or on-site configurable settings and the inspector needs to confirm that the installer has correctly set up such protective devices.

m Labelling of protective devices, switches and terminals

Sect 514
514.8.1 Each protective device must be arranged and identified so that the circuit protected
514.9.1 can be easily identified, and a diagram or chart indicating the function of each circuit and size of conductors is required; the inspector should have this key document in order to carry out much of his or her inspection and testing.

n Selection of equipment and protective measures appropriate to external influences

512.2
Sect 522 Equipment must be selected with regard to its suitability for the environment – ambient temperature, heat, water, foreign bodies, corrosion, impact, vibration, flora, fauna, radiation, building use and structure. A careful inspection is necessary to confirm the suitability of each item of equipment.

o Adequacy of access to switchgear and equipment

132.12
Sect 513 Every piece of equipment which requires operation or attention by a person must be so installed that adequate and safe means of access (related to the amount of its use) and sufficient working space are afforded; the inspector should check that these requirements are met.

p Presence of danger notices and other warning notices

Sect 514 Suitable warning notices, suitably located, are required to be installed to give warning of:

514.10 **Voltage**

▶ Where a nominal voltage exceeding 230 V to earth exists within an item of equipment or enclosure and where the presence of such a voltage would not normally be expected. The wording of this regulation was revised for Amendment No. 1 to BS 7671:2008 and it is clarified that only 'unusual' system voltages exceeding 230 volts to earth require warning labels. An example would be the use of a 690 volts three-phase a.c. power transformer used on an American air base located in the UK.

514.11 **Isolation**

▶ Where live parts are not capable of being isolated by a single device. The location of disconnectors should also be indicated except where there is no possibility of confusion.

Periodic inspection and testing

514.12.1 ▶ The wording of the notice is given in Regulation 514.12.1.

RCDs

514.12.2 ▶ The wording of the notice is given in Regulation 514.12.2.

Non-standard colours

514.14.1 ▶ For installations containing both pre-BS 7671:2008 (pre-harmonized colours) cable colours as well as cable colours to BS 7671:2008 and later editions (harmonized colours) an appropriate warning notice must be present at or near the relevant distribution board. The wording of the notice is given in Regulation 514.14.1.

Alternative supplies

514.15.1 ▶ For installations with alternative voltage sources, a 'multiple-source' warning notice is required at mains positions, points of isolation, distribution boards and at any remote metering. The wording of the notice is given in Regulation 514.15.1.

514.16 **High protective conductor current**
543.7.1.105 ▶ For circuits with a high protective conductor current, information must be provided at the relevant distribution board indicating these circuits, as required by Regulation 543.7.1.105.

Earthing and bonding connections

514.13.1 ▶ The requirements for the label and its wording are given in Regulation 514.13.1.

▶ The wording of the notice required when protection by earth-free local equipotential bonding (Regulation 418.2.5 refers) or by electrical separation for the supply to more than one item of equipment (Regulation 418.3 refers) is
514.13.2 given in Regulation 514.13.2.

q Erection methods

Chapter 52 contains detailed requirements on selection and erection. Fixings of switchgear, cables, conduit, fittings, etc. must be adequate for the environment and a detailed visual inspection is required during the erection stages as well as at completion.

2.6.3 Inspection checklist

Listed below are requirements to be checked when carrying out an installation inspection. The list is not exhaustive.

General

1 Equipment complies with a product standard or equivalent (511.1)
2 Equipment is installed using good workmanship (134.1.1)

3 Equipment is accessible for operation, inspection and maintenance (513.1)

4 Suitable for local atmosphere and ambient temperature (512.2). Installations in potentially explosive atmospheres are outside the scope of BS 7671

5 Final circuits are separate including the neutral conductors (314.4)

6 Protective devices identified to indicate the circuits they protect (514.8.1)

7 Protective devices adequate for intended purpose (Ch. 53)

8 Disconnection times likely to be met by installed protective devices (Ch. 41)

9 All circuits suitably identified (514.1, 514.8, 514.9)

10 Suitable main switch provided (Ch. 53)

11 Supplies to any safety services suitably installed, e.g. fire alarms to BS 5839 and emergency lighting to BS 5266

12 Means of isolation suitably labelled (514.1, 537.2.2.6)

13 Provision for disconnecting the neutral (537.2.1.7)

14 Main switches to single-phase installations, intended for use by an ordinary person, e.g. domestic, shop, office premises, to be double-pole (537.1.4)

15 RCDs provided where required (411.3.3, 411.5, 422.3.9, 522.6.102, 522.6.103, 532.1, 701.411.3.3, 702.53, 702.55.1, 702.55.4, 703.411.3.3, 704.411.3.2.1, 705.411.1, 705.422.7, 706.410.3.10, 708.553.1.13, 709.531.2, 710.411, 710.531.2.4, 711.410.3.4, 711.411.3.3, 712.411.3.2.1.2, 717.415.1, 721.411.1, 740.410.3, 740.415.1, 753.411.3.2, 753.415.1)

16 Discrimination between RCDs considered to avoid danger (314.1, 531.2.9)

17 Main earthing terminal provided (542.4.1), readily accessible and identified where separate from switchgear (514.13.1)

18 Provision for disconnecting earthing conductor (542.4.2)

19 Cables used comply with British or Harmonized Standards (Appendix 4 of BS 7671)

20 Earth tail pots installed where required on mineral insulated cables (543.2.7)

21 Non-conductive finishes on enclosures removed to ensure good electrical connection and if necessary made good after connecting (526.1)

22 Adequately rated distribution boards (to the relevant parts of BS EN 60439 or BS EN 61439 (may require derating, see GN 6.)

23 Correct fuses or circuit-breakers installed (531, 533)

24 All connections secure (134.1)

25 Consideration paid to overvoltage protection (443)

26 Consideration paid to electromagnetic disturbances (444)

27 Overcurrent protection provided where applicable (Ch. 43)

28 Suitable proximity (segregation) of circuits (528)

29 Label notice for first periodic inspection and test provided (514.12.1)

30 Sealing of the wiring system including fire barriers (527.2).

Switchgear

1 Meets requirements of BS EN 61008, BS EN 61009, BS EN 60947-2, BS EN 60898, relevant parts of BS EN 60439 or BS EN 61439 where applicable, or equivalent standards (511)

2 Securely fixed (134.1.1) and suitably labelled (514.1)

3 Non-conductive finishes on switchgear removed at protective conductor connections and if necessary made good after connecting (526.1)

4 Suitable cable glands and gland plates used (526.1)

5 Correctly earthed (Ch. 54)

6 Conditions likely to be encountered taken account of, i.e. suitable for the foreseen environment (512.2)

7 Suitable as means of isolation, where applicable (537.2.2)

8 Need for isolation, mechanical maintenance, emergency and functional switching met (537)

9 Firefighter's switch provided where required (537.6)

10 All connections secure (526)

11 Cables correctly terminated and identified (514, 526)

12 No sharp edges on cable entries, screw heads, etc., which could cause damage to cables (134.1.1, 522.8.11)

13 Adequate access and working space (132.12 and 513.1).

General wiring accessories

1 Complies with appropriate standards, for example, BS 5733 (general accessories) or BS EN 60670-22 (junction boxes) (511.1)

2 Box or other enclosure securely fixed (134.1.1)

3 Metal box or other enclosure earthed (Ch. 54)

4 Edge of flush boxes not projecting above wall surface (134.1.1)

5 No sharp edges on cable entries, screw heads, etc. which could cause damage to cables (134.1.1, 522.8.11)

6 Non-sheathed cables, and cores of cable from which sheath has been removed, not exposed outside the enclosure (526.8)

7 Conductors correctly identified (514.3)

8 Bare protective conductors having a cross-sectional area of 6 mm² or less to be sleeved green-and-yellow (514.4.2, 543.3.100)

9 Terminals tight and containing all strands of the conductors (526)

10 Cable grip correctly used or clips fitted to cables to prevent strain on the terminals (522.8.5, 526.6)

11 Adequate current rating (512.1.2).

Note: Reference should also be made to the recommendations contained in Approved Document M (England and Wales) and the Scottish Building Standards with regard to the heights at which socket-outlets, switches and other controls should be installed. See also the IET publication *Electrician's Guide to the Building Regulations*.

Lighting controls

1 Light switches comply with BS 3676 or BS EN 60669-1 (511.1)

2 Selected for external influences (512.2)

3 Single-pole switches connected in line conductors only (132.14.1)

4 Correct colour coding or marking of conductors (514.3)

5 Earthing of exposed metalwork, e.g. metal switchplate (Ch. 54)

6 Adequate current rating allowing for any capacitive or inductive effects (512.1.2)

7 Switch labelled to indicate purpose, where this is not obvious (514.1.1)

8 Appropriate controls suitable for the luminaires (559.6.1.9)

9 Standard wall accessory switches installed beyond zone 2 in bathroom.

Lighting points

1 Lights connected via a recognised accessory (559.6.1.1) with lampholders to BS EN 60598

2 Ceiling rose complies with BS 67 (559.6.1.1)

3 Luminaire couplers comply with BS 6972 or BS 7001 (559.6.1.1)

4 Track systems comply with BS EN 60570 (559.4.4)

5 Not more than one flex unless designed for multiple pendants (559.6.1.3)

6 Flex support devices used and suitable for the mass suspended (559.6.1.5)

7 Switch-lines identified (514.3.2 and Appendix 7 of BS 7671). For two-core switch wires, blue conductors are overmarked with brown or L at terminations; for three-, four- or five-core cables, all non-brown line conductors of switch and intermediate strappers are overmarked at terminations with brown or L

8 Penetrations in fire-rated ceiling made good (527.2.1)

9 Ceiling roses and similar not used for circuits having supply exceeding 250 V (559.6.1.2).

Socket-outlets

1 Comply with BS 546, BS 1363 or BS EN 60309-2 (553.1.3) and shuttered for household and similar installations (553.1.100)
2 Mounting height above the floor or working surface suitable (553.1.6)
3 Correct polarity (612.6)
4 If in a location containing a bath or shower, installed at least 3 m horizontally from the bath or shower unless shaver supply unit or SELV (701.512.3)
5 Protected where mounted in a floor (522)
6 Not used to supply a water heater having uninsulated elements (554.3.3)
7 Where metal conduit or earthed cable sheath or similar used as a protective conductor, presence of an earth tail between accessory box and socket-outlet terminal (543.2.7).

Junction boxes, joint box and terminations

1 All cable joints and terminations installed so that they are accessible for future inspection (except soldered, encapsulated, etc. joints or marked maintenance-free accessory; see 526.3)
2 Enclosures of terminals provide suitable protection against mechanical damage (526.7).

Fused connection unit

1 Correct rating and fuse (533.1)
2 Complies with BS 1363-4 (559.6.1.1 vii).

Cooker control unit

1 Sited to one side and low enough for accessibility and to prevent flexes trailing across radiant plates (522.2.1)
2 Cable to cooker fixed to prevent strain on connections (522.8.5).

Conduit systems

General

1 Securely fixed, box lids in place and adequately protected against mechanical damage (522.8)
2 Draw points are accessible (522.8.6)
3 Recommended quantity of cables for easy draw not exceeded during installation causing insulation damage; adequate boxes suitably spaced. Item should be inspected during the erection stage as the care and workmanship of the installer can be verified (522.8.1 and see *On-Site Guide* Appendix E)
4 Solid elbows and tees used only as appropriate (522.8.1)
5 Unused entries blanked off where necessary (416.2)
6 Conduit system components comply with a relevant British Standard (511.1)
7 Provided with drainage holes and gaskets as necessary (522.3)
8 Radius of bends such that cables are not damaged (522.8.3).

Rigid metal conduit

1 Complies with BS EN 61386-21 (521.6)
2 Connected to the main earthing terminal (411.4.2, 411.5.1)
3 Line, neutral and any additional protective conductors are contained in the same conduit (521.5.1)
4 Conduit suitable for wet, damp or corrosive situations (522.3, 522.5)
5 Fixed appropriately (522.8 and see Guidance Note 1 Appendix G)
6 Unpainted ends and scratches, etc. protected by painting (134.1.1, 522.5)
7 Ends of conduit reamed and bushed where relevant (134.1.1, 522.8).

Rigid non-metallic conduit

1   Complies with BS 4607, BS EN 60423 or the BS EN 61386 series (521.6)
2   Ambient and working temperatures within permitted limits (522.1, 522.2)
3   Provision made to allow for expansion and contraction
4   Boxes and fixings suitable for mass of luminaire suspended at expected temperature (559.6.1.5).

Flexible metal conduit

1   Complies with BS EN 61386 series (521.6)
2   Separate protective conductor provided (543.2.3)
3   Adequately supported and terminated (522.8)
4   Line, neutral and any additional protective conductors are contained in the same conduit (521.5.1).

Trunking

General

1   Complies with BS 4678 or BS EN 50085-1 (521.6)
2   Securely fixed and adequately protected against mechanical damage (522.8)
3   Selected, erected and routed to avoid ingress of water (522.3)
4   Proximity to non-electrical services, i.e. sources of heat, smoke etc. cannot cause damage (528.3)
5   Internal sealing provided where necessary (requires inspection during the erection stage) (527.2.2)
6   Holes surrounding trunking made good (527.2.1)
7   Band I circuits partitioned from Band II circuits or insulated for the highest voltage present (528.1)
8   Circuits partitioned from Band I circuits or wired in mineral insulated metal-sheathed cables (528.1)
9   Cables supported for vertical runs (522.8.5).

Metal trunking

1   Line, neutral and any additional protective conductors are contained in the same trunking (521.5.1)
2   Protected against damp or corrosion (522.3, 522.5)
3   Earthed (411.4.2, 411.5.1)
4   Joints mechanically sound and of adequate continuity (543.2.5).

Busbar trunking and powertrack systems

1   Busbar trunking system complies with BS EN 60439-2 or BS EN 61439-6 or other appropriate standard; powertrack system complies with BS EN 61534 series or other appropriate standard (521.4)
2   Securely fixed and adequately protected against mechanical damage (522.8)
3   Joints mechanically sound and of adequate continuity (543.2.5).

Insulated cables

Non-flexible cables

1   Correct type (521)
2   Correct current rating (523)
3   Protected against mechanical damage and abrasion (522.8)
4   Cables suitable for high or low ambient temperature as necessary (522.1)
5   Non-sheathed cables protected by enclosure in conduit, duct or trunking (except for protective conductors of 4 mm² and larger 521.10)
6   Sheathed cables

- ▶ routed in allowed zones or mechanical protection provided (522.6.101)
- ▶ in the case of domestic or similar installations not under the supervision of skilled or instructed persons, additional protection is provided by RCD having $I_{\Delta n}$ not exceeding 30mA (522.6.102)

7 Cables in partitions containing metallic structural parts in domestic or similar installations not under the supervision of skilled or instructed persons provided with

- ▶ adequate mechanical protection to suit both the installation of the cable and its normal use
- ▶ additional protection by RCD having $I_{\Delta n}$ not exceeding 30 mA (522.6.103)

8 Where exposed to direct sunlight, of a suitable type or suitably shaded (522.11)

9 Not run in lift shaft unless part of the lift installation and of the permitted type (528.3.5)

10 Buried cable correctly selected and installed for use (522.8.10)

11 Correctly selected and installed for use overhead (522.8.4)

12 Internal radii of bends not sufficiently tight as to cause damage to cables or to place undue stress on terminations to which they are connected (522.8.3 and Guidance Note 1 Appendix G)

13 Correctly supported (522.8.4, 522.8.5)

14 Not exposed to water, etc. unless suitable for such exposure (522.3)

15 Metal sheaths and armour earthed (411.3.1.1)

16 Identified at terminations (514.3)

17 Joints and connections electrically and mechanically sound and adequately insulated (526.1, 526.2)

18 All wires securely contained in terminals, etc. without strain (522.8.5, 526)

19 Enclosure of terminals (526)

20 Glands correctly selected and fitted with shrouds and supplementary earth tags as necessary (526.1)

21 Joints and connections mechanically sound and accessible for inspection, except as permitted otherwise (526.1, 526.3).

Flexible cables (521.9)

1 Correct type (521.9.1)

2 Correct current rating (523)

3 Protected where exposed to mechanical damage (522.6, 522.8)

4 Suitably sheathed where exposed to contact with water (522.3) or corrosive substances (522.5)

5 Protected where used for final connections to fixed apparatus, etc. (526.8)

6 Selected for resistance to damage by external heat sources (522.2)

7 Segregation of Band I and Band II circuits (528; see also BS 6701)

8 Fire alarm and emergency lighting circuits segregated (528; see also BS 5839, BS 5266)

9 Cores correctly identified (514.3)

10 Connections to have durable electrical continuity, adequate mechanical strength and be made using appropriate means (526.1, 526.2)

11 Where used as fixed wiring, relevant requirements met (521.9.1)

12 Final connections to current-using equipment properly secured and arranged to prevent strain on connections (526.6)

13 Mass supported by cable to not impair safety of the installation (559.6.1.5).

Protective conductors

1 Cables incorporating protective conductors comply with the relevant BS (511.1)

2 Joints in metal conduit, ducting or trunking comply with Regulations (543.3)

3 Flexible or pliable conduit is supplemented by a protective conductor (543.2.3)

4 Minimum cross-sectional area of copper conductors (543.1)

5 Copper conductors of 6 mm² or less protected by insulation (543.3.100)

6 Circuit protective conductor at termination of sheathed cables insulated with sleeving (543.3.100)

7 Bare circuit protective conductor protected against mechanical damage and corrosion (543.3.1)

8 Insulation, sleeving and terminations identified by colour combination green-and-yellow (514.4.2)

9 Joints electrically and mechanically sound (526.1)

10 Separate circuit protective conductors of not less than 4 mm² if not protected against mechanical damage (543.1.1)

11 Main and supplementary bonding conductors of correct size (544).

Enclosures
General

1 Suitable degree of protection ('IP Code' in BS EN 60529) appropriate to external influences (416.2, 522, Part 7).

2.7 Initial testing

The test methods described in this section are recommended to be used for verification. In this respect Guidance Note 3 is cited in Regulation 612.1. This does not rule out the use of other test methods provided they give valid results.

612.1

2.7.1 Test results

The test results must be recorded on the Schedule(s) of Test Results and compared with relevant criteria. For example, in order to verify disconnection times, the relevant criteria would be design earth fault loop impedance values provided by the designer.

A model Schedule of Test Results is shown in Chapter 5.

2.7.2 Electrical Installation Certificate

631.1 Regulation 631.1 of BS 7671 requires that, upon completion of the verification of a new, modified or extended installation, an Electrical Installation Certificate based on the model given in Appendix 6 of BS 7671 shall be provided. Chapter 63 requires that:

632.1 **1** the Electrical Installation Certificate be accompanied by schedules of inspection and schedules of test results. These schedules shall be based upon the models given in Appendix 6 of BS 7671

632.2 **2** the schedule of test results shall include test results for every circuit

631.4 **3** the Electrical Installation Certificate is signed or otherwise authenticated by the competent person responsible for each facet of design, construction and inspection and test

Note: The person responsible for carrying out the initial verification and signing the inspection and test box has certain responsibilities for checking some design and construction aspects (see Regulation 611.2, Section 632 and section 2.1 of this Guidance Note).

632.4 **4** any defects or omissions revealed by the inspector shall be made good, and as necessary inspected and tested again, before the Electrical Installation Certificate is issued; it is not the responsibility of the person or organization carrying out the inspection and testing to make good defects or omissions.

2.7.3 Model forms

Typical forms for use when carrying out inspection and testing are included in Chapter 5 of this Guidance Note.

612.1 ### 2.7.4 The sequence of tests

Initial tests should be carried out in the following sequence where relevant and practical:

a Continuity of protective conductors, including main and supplementary bonding (2.7.5);

b Continuity of ring final circuit conductors (2.7.6);

c Insulation resistance (2.7.7);

d Protection by SELV, PELV or by electrical separation (2.7.8, 2.7.9);

e Protection by barriers or enclosures provided during erection (2.7.10);

f Insulation resistance of non-conducting floors and walls (2.7.11);

g Polarity (2.7.12);

h Earth electrode resistance (2.7.13);

i Protection by automatic disconnection of the supply (2.7.14);

j Earth fault loop impedance (2.7.15);

k Additional protection (2.7.18);

l Prospective fault current (2.7.16);

m Phase sequence (2.7.17);

n Functional testing (2.7.18, 2.7.19);

o Voltage drop (2.7.20).

612.2.1 ### 2.7.5 Continuity of protective conductors, including main and supplementary bonding

411.3.1.1 Regulation 411.3.1.1 requires that installations which provide protection against electric shock using automatic disconnection of supply must have a circuit protective conductor run to and terminated at each point in the wiring and at each accessory. An exception is made for a lampholder having no exposed-conductive-parts and suspended from such a point.

As such, Regulation 612.2.1 requires that a continuity check be carried out on all circuits including ring circuits.

There are two widely used test methods that have evolved for checking conductor continuity. 'Test method 1' uses the circuit cable shorted out and 'Test method 2' uses a supplementary length of test cable (this method being popularly known as the 'wandering lead' method).

Instrument: Use a low resistance ohmmeter for these tests – see section 4.3.

Every protective conductor, including circuit protective conductors, the earthing conductor and main and supplementary equipotential bonding conductors, should be tested to verify that the conductors are electrically sound and correctly connected.

Test method 1 detailed below, as well as checking the continuity of the protective conductor, also measures $(R_1 + R_2)$ which, when added to the external impedance (Z_e), enables the earth fault loop impedance (Z_s) to be checked against the design (see 2.7.15). $(R_1 + R_2)$ is the sum of the resistances of the line conductor R_1 and the circuit protective conductor R_2.

Test readings may be affected by parallel paths through exposed-conductive-parts and/or extraneous-conductive-parts.

Parallel earth paths and effects on test readings

Inspectors should be aware of the possible existence of parallel earth return paths. These may take the form of metallic cable management products, extraneous-conductive-parts or indeed other metallic parts. Examples include installations incorporating steel conduit, steel trunking, micc, steel wire armour or other metal sheathed cables, equipment and accessory boxes fitted to metal stud walls or to the building fabric, and luminaires fitted in grid ceilings. They exist in domestic, commercial and industrial installations.

The effect of parallel earth return paths is that the measured value of the continuity, R_2, tends towards a zero value. It is often impractical and in some cases impossible to carry out testing with some or all of the parallel paths disconnected and the inspector must be aware of this.

Test method 1 (for circuits)

Make a temporary cable shorting link and connect the line conductor to the protective conductor at the distribution board or consumer unit. Then test between line and earth terminals at each outlet in the circuit as shown in Figure 2.1a. The resistance of the test leads should either be auto-nulled by the test instrument or measured and deducted from the readings obtained.

Where the installation has all-insulated wiring (see notes on parallel earth paths and effects on test results above) and the cable accessories are not in contact with earth, then this test measures $(R_1 + R_2)$, i.e. the resistance of the line conductor, R_1, plus the resistance of the protective conductor, R_2, for that circuit which, if added to the earth fault loop impedance at the distribution board, can be taken as the circuit's earth fault loop impedance. It is important to record the value of $(R_1 + R_2)$ obtained at the circuit's extremity, namely the furthest circuit distance from the distribution board.

▼ **Figure 2.1a**

Connections for testing continuity of protective conductors: method 1

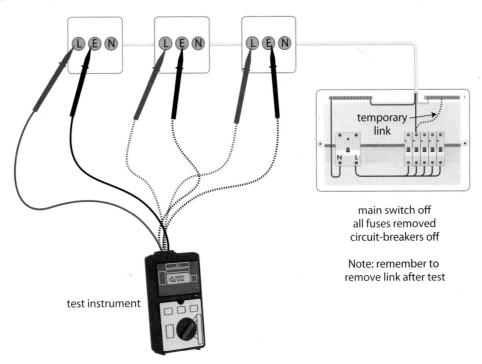

temporary link

main switch off
all fuses removed
circuit-breakers off

Note: remember to remove link after test

test instrument

Expected results for test method 1

The results should first and foremost indicate no open circuit in the protective conductors. For insulated wiring systems installed in conditions where accessory boxes and similar are not connected to fabric or other elements that may be earthed, then as stated earlier the readings measured will be the sum of the line and protective conductor resistances, or $(R_1 + R_2)$. This test can detect poor continuity at junctions and connections since, for a new installation with new accessories, the contribution of resistance of healthy connections to the measured resistance is negligible and can be ignored. Thus, by employing the resistance data for copper conductors given in Appendix B expected values for healthy circuits can be approximated, and compared with the test readings obtained.

As an example, a radial circuit of length about 55 m with 2.5 mm² line and protective conductors should have an $(R_1 + R_2)$ resistance as follows:

Length of circuit is 55 m

Resistance of cable is 7.41 mΩ/m (at 20°C) from Table B1

Theoretical minimum d.c. resistance = (55 × 2 × 7.41)/1000 = 0.82 ohm

When verifying this circuit the inspector should be looking for a reading of that order, so a reading of, say, 0.8 to 1.2 ohms would be acceptable. If the circuit had several outlets, thus meaning that the circuit conductors were broken and connected in screw terminals at each accessory, then a slightly higher value may be measured, as there would be some resistance at the terminations.

Test method 2 (for circuits)
Instrument: Use a low-resistance ohmmeter for this test. Refer to section 4.3.

One lead of the test instrument is connected to the earth terminal at the distribution board via a length of test cable or 'wandering lead'. The other test lead is used to make contact with the protective conductor at various points on the circuit under test, e.g. luminaires, switches, spur outlets, etc. as shown in Figure 2.1b. The resistance of the wandering lead and the test leads are either auto-nulled prior to making the test or measured and subtracted from measured readings.

This test measures the continuity resistance of the circuit protective conductor, R_2, which should be recorded on the Schedule of Test Results (see earlier note 'Parallel earth paths and effects on test readings').

▼ **Figure 2.1b**
Connections for testing continuity of protective conductors: method 2

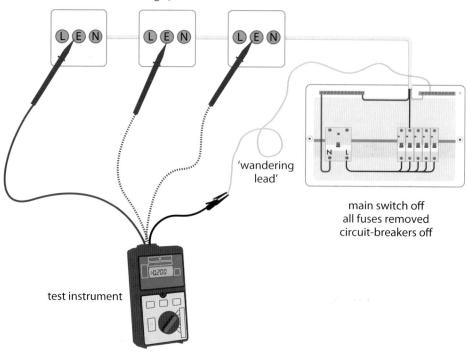

'wandering lead'

main switch off
all fuses removed
circuit-breakers off

test instrument

Expected results for test method 2

The results should first and foremost indicate no open circuit in the protective conductors. For insulated wiring systems installed in conditions where accessory boxes and similar are not connected to fabric or other elements that may be earthed, then the measurement will equate to the protective conductor resistance R_2. This test can detect poor continuity at junctions and connections since, for a new installation with new accessories, the contribution of resistance of healthy connections is negligible and can be ignored. Thus, by employing the resistance data for copper conductors given in Appendix B expected values for healthy cable and connections can be checked.

As an example, a radial circuit of length about 55 m with a 2.5 mm² copper protective conductor should have an R_2 resistance as follows:

Length of circuit is 55 m

Resistance of cable is 7.41 mΩ/m (at 20°C) from Table B1

Theoretical minimum d.c. resistance = (55 × 7.41)/1000 = 0.41 ohm

When verifying this circuit the inspector should be looking for a reading of that order, say 0.4 to 0.5 ohm. If the circuit had several outlets, thus meaning that the protective conductor was broken and connected in screw terminals at each accessory, then a slightly higher value may be measured, as there would be some resistance at the terminations.

Testing bonding conductors and earthing conductors

To confirm the continuity of these protective conductors, **test method 2** may be used.

This method can also be used to confirm a bonding connection between extraneous-conductive-parts where it is not possible to see a bonding connection, e.g. where bonding clamps have been 'built in'. The test would be done by connecting the leads of the instrument between any two points such as metallic pipes and looking for a low reading of the order of 0.05 ohm (it should be noted that not all low-resistance ohmmeters can read this low, see section 4.3).

Where metallic enclosures have been used as the protective conductors, e.g. conduit, trunking, steel-wire armouring, etc. the following procedure should be employed:

1 Inspect the enclosure along its length for soundness of construction
2 Perform the standard continuity test using the appropriate test method described above.

Instrument: Use a low-resistance ohmmeter for this test – section 4.3.

Expected test results

The results should first and foremost indicate no open circuit in the protective conductors. For lengths of conductor use Appendix B for resistance data. For joints across bonds by earth clamps and similar, the readings should approach 0.05 ohm taking into account both the resolution of the instrument and its accuracy at low values.

2.7.6 Continuity of ring final circuit conductors

612.2.2 A three-step test is required to verify the continuity of the line, neutral and protective conductors and correct wiring of every ring final circuit. The test results show if the ring has been interconnected to create an apparently continuous ring circuit which is in fact broken or connected as a 'figure of eight' configuration.

Instrument: Use a low-resistance ohmmeter for this test – see section 4.3.

Step 1
The line, neutral and protective conductors are visually identified at the distribution board or consumer unit and the end-to-end resistance of each is measured separately (see Figure 2.2a)

▼ **Figure 2.2a**
Connections for testing
step 1

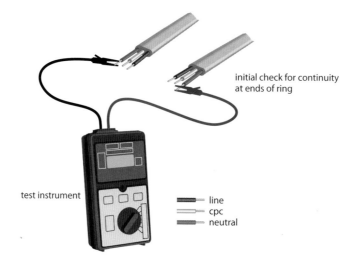

initial check for continuity at ends of ring

test instrument

line
cpc
neutral

These resistances are r_1, r_n and r_2 respectively. A finite reading confirms that there is no open circuit on the ring conductors under test. The resistance values obtained should be of the same order if the conductors are the same size. If the protective conductor has a smaller cross-sectional area, the resistance r_2 of the protective conductor loop will be proportionally higher than that of the line or neutral loop, e.g. 1.67 times for 2.5/1.5 mm² cable. If the resistance readings are not as expected this could mean the following:

▶ *readings lower than the expected resistance*[1], would suggest that the ring is incorrectly configured, possibly wired in a 'figure of eight' connection; this may be further confirmed by the step 2 test below
▶ *readings higher than the expected resistance*[1], would suggest that one or more of the conductor terminations is poorly made

[1] The 'expected resistance' mentioned above is that found from the tabulated d.c. resistance for the conductor size per metre multiplied by the installed length and corrected for measured temperature. A small allowance should be made for instrument errors. Table B1 gives values of d.c. resistance for conductors.

Step 2
The open ends of the line and neutral conductors are then connected together so that the outgoing line conductor is connected to the returning neutral conductor and vice versa (see Figure 2.2b).

▼ Figure 2.2b
Connections for testing step 2

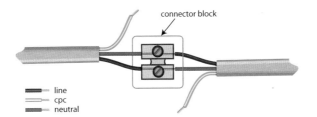

The resistance between line and neutral conductors is measured at each socket-outlet. The readings at each of the sockets wired into the ring should be substantially the same and the value will be approximately one-quarter of the resistance of the line plus the neutral loop resistances, i.e. $(r_1 + r_n)/4$ (see mathematical explanation in Figure 2.3). Any sockets wired as spurs will give a higher resistance value due to the resistance of the spur conductors.

Note: Where single-core cables are used, care should be taken to verify that the line and neutral conductors of **opposite** ends of the ring circuit are connected together. An error in this respect will be apparent from the readings taken at the socket-outlets, progressively increasing in value as readings are taken towards the midpoint of the ring, then decreasing again towards the other end of the ring.

Step 3
The open ends of the line conductor and cpc are then cross-connected (see Figure 2.2c).

▼ Figure 2.2c
Connections for testing step 3

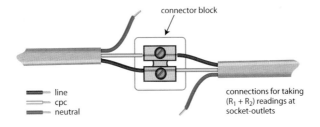

The resistance between line and earth is measured at each socket-outlet. The readings obtained at each of the sockets wired into the ring will be substantially the same and the value will be approximately one-quarter of the resistance of the line plus cpc loop resistances, i.e. $(r_1 + r_2)/4$ (the explanation for this being similar to step 2). A higher resistance value will be recorded at any sockets wired as spurs. The highest value recorded represents the maximum $(R_1 + R_2)$ of the circuit and is recorded on the Schedule of Test Results. The value can be used to determine the earth fault loop impedance (Z_s) of the circuit to verify compliance with the loop impedance requirements of the Regulations (see section 2.7.14).

The inspector is again reminded to take note of the effects of possible parallel return paths on these continuity tests, described in 2.7.5.

CITY OF LIVERPOOL COLLEGE
VAUXHALL ROAD
LIVERPOOL
L3 6BN

▼ **Figure 2.3** Explanation of the maths for step 2

Figures 2.3a to e explain the expected results for a correctly wired ring circuit.

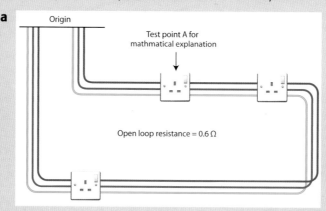

Figure **a** above is an example of a correctly wired ring, the open loop resistances from step 1 being 0.6 ohm. A test point about a third distance round the ring is used to illustrate the maths as explained below.

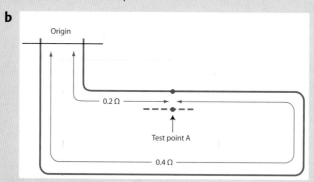

Figures **b** and **c** show the resistances of each leg of the ring as a test is applied at this point as per step 2 (line–neutral).

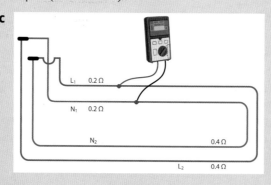

The equivalent connects are then represented in figure **d**.

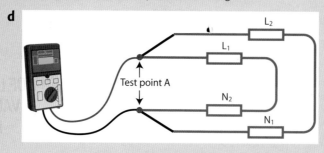

The equivalent circuit diagram and resultant resistance are shown in figure **e**.

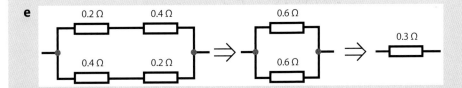

Thus, in summary the open loop resistances are 0.6 ohm for both line and neutral, giving an $(r_1 + r_n)$ value of $(0.6 + 0.6)$, or 1.2 ohms.

From Figure **e** above it can be seen that a correctly connected ring will give a step 2 reading of a quarter of the $(r_1 + r_n)$ value, or:

$$R_{step\ 2\ test} = \frac{(r_1 + r_n)}{4}, \text{ in this case } \frac{1.2}{4} = 0.3 \text{ ohm}$$

612.3

2.7.7 Insulation resistance

Insulation resistance testing is a fundamental test for inspectors. Often on larger construction sites, cables will be insulation resistance tested during various stages of installation to prove the integrity of installed cables. It is always preferred to re-test cables and equipment for insulation resistance as part of initial verification as well as during construction.

612.3.1
BS 7671 requires that insulation resistance is measured between all of the live conductors and between the live conductors and the protective conductor with the protective conductor connected to the earthing arrangement. This key change to the procedure was introduced in the 17th Edition in 2008 and is an important change to practice for many installers and inspectors. Taking cables as an example, previously it was acceptable to test a cable between the various cores, and test to earth (often the armouring or sheath of the cable); sometimes these cables were terminated without further testing. This is not acceptable now and it is essential to test to the armouring with it connected (with a fly-lead if necessary) to the installation earthing arrangement. This is shown in Figure 2.4c. It is a good idea to test all cables including those tested during the construction stage using this method.

The purpose of the insulation resistance test is to verify that the insulation of conductors provides adequate insulation, is not damaged and that live conductors or protective conductors are not short-circuited.

As a reminder, prior to carrying out the test, check that:

1 the protective conductor of the item (switchgear or cable etc.) is connected to the main earthing terminal
2 pilot or indicator lamps, and capacitors are disconnected from circuits to avoid an inaccurate test value being obtained (see note below)
3 voltage-sensitive electronic equipment such as dimmer switches, touch switches, delay timers, power controllers, electronic starters for fluorescent lamps, emergency lighting, RCDs and similar equipment are disconnected so that they are not subjected to the test voltage.

Note: 2 and 3 are necessary due to the fact that as testing occurs between all conductors, anything connected and in circuit will be subjected to the test voltage.

Instrument: Use an insulation resistance tester – see section 4.4.

Table 61   Insulation resistance tests should be carried out using the appropriate d.c. test voltage specified in Table 61 of BS 7671. The installation will be deemed to conform with the Regulations in this respect if the main switchboard, and each distribution circuit tested separately with all its final circuits connected, but with current-using equipment disconnected, has an insulation resistance not less than that specified in Table 61, reproduced here as Table 2.2.

▼ **Table 2.2** Minimum values of insulation resistance

Circuit nominal voltage	Test voltage d.c. (V)	Minimum insulation resistance (MΩ)
SELV and PELV	250	0.5
Up to and including 500 V with the exception of SELV and PELV but including FELV	500	1.0
Above 500 V	1000	1.0

Simple installations that contain no distribution circuits should preferably be tested as a whole, see example in Figure 2.4a.

The tests should be carried out with the main switch off, all fuses in place, switches and circuit-breakers closed, lamps removed, and fluorescent and discharge luminaires and other equipment disconnected. Where the removal of lamps and/or the disconnection of current-using equipment is impracticable, the local switches controlling such lamps and/or equipment should be open.

To perform the test in a complex installation it may need to be subdivided into its component parts.

Although an insulation resistance value of not less than 1 megohm complies with the Regulations new installations should not yield test results this low.

Example (i) – Insulation resistance test of a whole consumer unit

▼ **Figure 2.4a**
Example of an insulation resistance test of a whole consumer unit

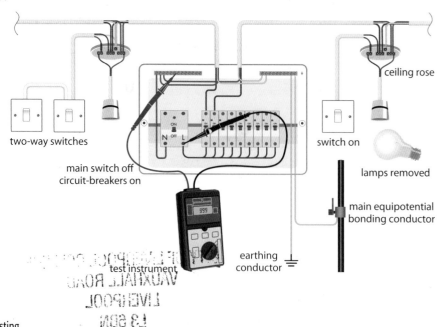

Figure 2.4a shows an example of testing a whole consumer unit (i.e. installation) in one test (only the line to neutral test is shown). The tests required are a test between the live conductors (line to neutral) and tests between the live conductors and earth (line to earth and neutral to earth).

For circuits containing two-way switching or two-way and intermediate switching, the switches must be operated one at a time and the circuits subjected to additional insulation resistance test in these configurations.

For circuits/equipment vulnerable to the test voltage, the test is made with the line and neutral conductors connected together and earth. It is essential that the incoming earth connection is connected to the installation main earthing terminal (and that this is connected to the means of earthing) for these tests.

Example (ii) – Insulation resistance test of a final circuit

Figure 2.4b shows an example of testing a single final circuit at a consumer unit (only the line to neutral test is shown). The tests required are a test between the live conductors (line to neutral) and tests between the live conductors and earth (line to earth and neutral to earth).

For circuits containing two-way switching or two-way and intermediate switching, the switches must be operated one at a time and the circuits subjected to additional insulation resistance test in these configurations.

For circuits/equipment vulnerable to the test voltage, the test is made with the line and neutral conductors connected together and earth. It is essential that the incoming earth connection is connected to the installation main earthing terminal (and that this is connected to the means of earthing) for these tests.

▼ **Figure 2.4b**
Example of insulation resistance test of a final circuit

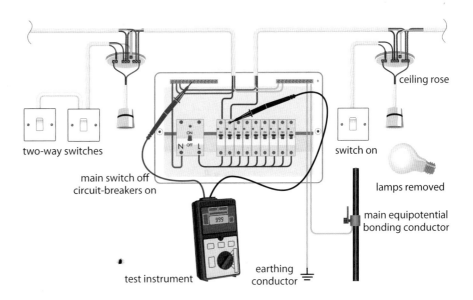

ceiling rose

two-way switches

main switch off
circuit-breakers on

switch on

lamps removed

main equipotential
bonding conductor

test instrument

earthing
conductor

Note 1: the test should be initially carried out on the complete installation
Note 2: bonding and Earthing connections are in place

CITY OF LIVERPOOL COLLEGE
VAUXHALL ROAD
LIVERPOOL
L3 6BN

Insulation resistance testing of a three-phase 4-core power cable

The cable is tested as per Table 2.3.

▼ **Table 2.3** Insulation resistance test on 4-core power cable

Test 1	L_1 to L_2	
Test 2	L_1 to L_3	The lowest value of these tests is recorded as 'between live conductors'
Test 3	L_2 to L_3	
Test 4	$L_1 + L_2 + L_3$ (connected together) to neutral	
Test 5	$L_1 + L_2 + L_3$ (connected together) to earth	The lowest value of these tests is recorded as 'between live conductors and earth'
Test 6	neutral to earth	

Note: It is essential for tests 5 and 6 that the cable earth is connected to the installation earthing terminal.

▼ **Figure 2.4c**
Insulation testing of a three-phase power cable (showing the neutral to earth test)

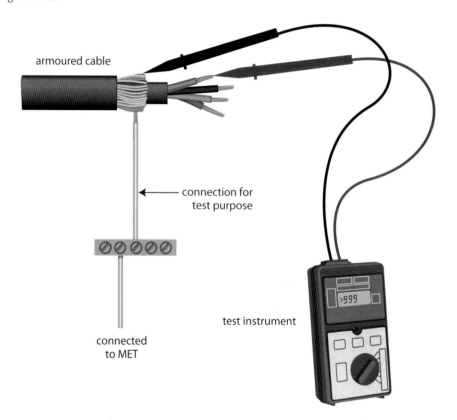

armoured cable

connection for test purpose

connected to MET

test instrument

Insulation resistance readings obtained should be not less than the minimum values referred to in Table 2.2.

2.7.8 Confirming SELV or PELV circuits by insulation testing

612.4

In order to establish which insulation tests are required for verifying a SELV or PELV system, the requirements of Section 414 of BS 7671 must firstly be understood.

Sect 414

There are situations where the provision of insulation of SELV or PELV circuits for basic protection is generally not required by BS 7671, i.e. for the following voltages:

414.4.5 ▶ up to 12 volts a.c. or 30 volts d.c. in wet areas

▶ up to 25 volts a.c. or 60 volts d.c. in dry areas.

Part 7 However, for bathrooms, swimming pools, saunas and some other special locations basic protection by insulation is required for SELV and PELV at all voltages.

It is often, therefore, easier to carry out insulation tests of these circuits as a matter of course.

Where SELV or PELV is used as a protective measure and insulation testing is required, Tables 2.4 and 2.5 set out the requirements.

Instrument: Use an insulation resistance tester for these tests. Refer to section 4.4.

612.4.1 ▼ **Table 2.4** SELV insulation resistance tests

Test type	Description	Test voltage d.c. (V)	Minimum acceptable resistance (MΩ)
Basic insulation	Between line conductors and all other circuits including other SELV and PELV and low voltage circuits	250	0.5
Line to Earth	Between all SELV live parts and Earth	250	0.5

Note: In situations where the SELV conductors are separated by just insulation, such as within a multicore cable with low voltage circuits, then the test voltage shall be increased to 500 volts d.c. and the insulation resistance shall be not less than 1 MΩ.

612.4.2 ▼ **Table 2.5** PELV insulation resistance tests

Test type	Description	Test voltage d.c. (V)	Minimum acceptable resistance (MΩ)
Basic insulation	Between line conductors and all other circuits including other SELV and PELV and low voltage circuits	250	0.5

Note: In situations where the PELV conductors are separated by just insulation, such as within a multicore cable with low voltage circuits, then the test voltage shall be increased to 500 volts d.c. and the insulation resistance shall be not less than 1 MΩ.

612.4.3 ## 2.7.9 Testing of electrically separated circuits

The source of supply should be inspected to confirm compliance with the Regulations.
413.3.2 In addition, should any doubt exist, the voltage should be measured to verify it does not exceed 500 V.

Insulation tests are then made as per Table 2.6.

Instrument: Use an insulation resistance tester for these tests. Refer to section 4.4.

▼ Table 2.6 Tests made to verify electrical separation

Test type	Description	Test voltage d.c. (V)	Minimum acceptable resistance (MΩ)
Basic separation	Between the electrically separated live conductors and the isolating transformer supply live conductors	500	1
Basic insulation of the separated conductors	Between the electrically separated live conductors and their corresponding exposed-conductive-parts	500	1
Basic insulation of any exposed-conductive-parts associated with separated conductors	Between any exposed-conductive-parts associated with the electrically separated circuits and any protective conductor, other exposed-conductive-parts or Earth	500	1

Additional inspections and tests for separated circuit supplying more than one item of current-using equipment:

418.3
1 Apply a continuity test between all exposed-conductive-parts of the separated circuit to ensure that they are bonded together. This equipotential bonding should then be subjected to a 500 V d.c. insulation resistance test between it and the protective conductor or exposed-conductive-parts of other circuits, or to extraneous-conductive-parts. The insulation resistance should be not less than 1.0 MΩ.
Instruments: Use a low-resistance ohmmeter and an insulation resistance tester for these tests. Refer to Chapter 4.

2 All socket-outlets must be inspected to ensure that the protective conductor contact is connected to the equipotential bonding conductor.

3 All flexible cables other than those feeding Class II equipment must be inspected to ensure that they contain a protective conductor for use as an equipotential bonding conductor.

418.3.7
Table 41.2
Table 41.3

Table 41.1
4 Operation of the protective device must be verified by measurement of the fault loop impedances (i.e. between live conductors) to the various items of connected equipment. These values can then be compared with the relevant maximum Z_s values given by Regulations 411.4.6 for fuses and 411.4.7 for circuit-breakers (Tables 41.2 and 41.3 for 230 V systems), with reference to the type and rating of the protective device for the separated circuit. Although these tables pertain to the line/protective conductor loop path, and the measured values are between live conductors, they give a reasonable approximation to the values required to achieve the required disconnection time of Table 41.1.

612.4.5
2.7.10 Protection by barriers or enclosures provided during erection

This test is not applicable to barriers or enclosures of factory-built equipment. It is applicable to those constructed on site during the course of assembly or erection and therefore is seldom necessary. Where, during erection, an enclosure or barrier is provided for basic protection, a degree of protection not less than IPXXB or IP2X is required. Readily accessible horizontal top surfaces must have a degree of protection of at least IPXXD or IP4X.

416.2.1
416.2.2
416.2.1

Whilst enclosures are covered by product standards, barriers may not be and the inspector must use judgement in deciding if a barrier is fit for purpose.

The degree of protection afforded by IP2X is defined in BS EN 60529 as protection against the entry of 'Fingers or similar objects not exceeding 80 mm in length. Solid objects exceeding 12.5 mm in diameter'. The test is made with a metallic standard test finger (test finger 1 to BS EN 61032).

Both joints of the finger may be bent through 90 ° with respect to the axis of the finger, but in one and the same direction only. The finger is pushed without undue force (not more than 10 N) against any openings in the enclosure and, if it enters, it is placed in every possible position.

A SELV supply, not exceeding 50 V, in series with a suitable lamp is connected between the test finger and the live parts inside the enclosure. Conducting parts covered only with varnish or paint, or protected by oxidation or by a similar process, must be covered with a metal foil electrically connected to those parts which are normally live in service.

The protection is satisfactory if the lamp does not light.

The degree of protection afforded by IP4X is defined in BS EN 60529 as protection against the entry of 'Wires or strips of thickness greater than 1.0 mm, and solid objects of 1.0 mm diameter or greater'.

The test is made with a straight rigid steel wire of 1 mm diameter applied with a force of 1 N ± 10 per cent. The end of the wire must be free from burrs, and at a right angle to its length.

The protection is satisfactory if the wire cannot enter the enclosure.

Reference should be made to the appropriate product standard or BS EN 60529 for a fuller description of the degrees of protection, details of the standard test finger and other aspects of the tests.

418.1 **2.7.11 Proving and testing of non-conducting location (insulation resistance/impedance of floors and walls)**

Where fault protection is provided by a non-conducting location, the following should be verified, prior to carrying out insulation testing:

418.1.2 **1** Exposed-conductive-parts should be inspected to confirm that no one can come into simultaneous contact with:
  ▶ two exposed-conductive-parts, or
  ▶ an exposed-conductive-part and any extraneous-conductive-part

418.1.3 **2** In a non-conducting location there must be no protective conductors
  3 Any socket-outlets installed in the location must not incorporate an earthing contact.

418.1.5 Following these checks, the insulation resistance between the insulating floors and walls to the installation main earthing terminal (via a local earth terminal of the general
612.5.1 installation) should be measured. It is required that at least three measurements are made. One measurement must be made approximately one metre from any accessible extraneous-conductive-part, e.g. metal pipe, in the location and the other measurements should be made at distances further away. Methods of measuring the insulation resistance/impedance of floors and walls are described below.

Test method

The insulation resistance test may be made using an insulation resistance tester, see section 4.4, and the test is made between test electrode 1 or test electrode 2 (see Figures 2.5a and 2.5b) and the main protective conductor of the installation.

Appx 13   ### Measuring insulation resistance of floors and walls

A magneto-ohmmeter or battery-powered insulation resistance tester providing a no-load voltage of approximately 500 V (or 1000 V if the rated voltage of the installation exceeds 500 V) is used as a d.c. source.

The resistance is measured between the test electrode and the main protective conductor of the installation.

The test electrodes may be either of the following types. In case of dispute, the use of test electrode 1 is the reference method.

It is recommended that the test be made before the application of the surface treatment (varnishes, paints and similar products).

Test electrode 1

The test electrode shown in Figure 2.5a comprises a metallic tripod of which the parts resting on the floor form the points of an equilateral triangle. Each supporting part is provided with a flexible base ensuring, when loaded, close contact with the surface being tested over an area of approximately 900 mm² and having a combined resistance of less than 5000 Ω between the terminal and the conductive rubber pads.

Before measurements are made, the surface being tested is cleaned with a cleaning liquid. While measurements of the floors and walls are being made, a force of approximately 750 N (75 kg in weight) for floors or 250 N for walls, is applied to the tripod.

▼ **Figure 2.5a**
Test electrode 1

Fig 13A

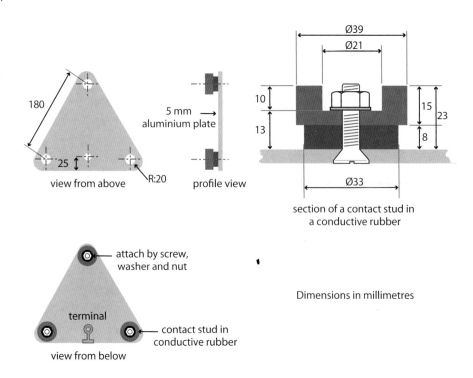

Dimensions in millimetres

ECHLIOO TOOARBAIT 70 YTIO
GAOR JJAHXUAV
JOOARBAIJ
M82 E I

Test electrode 2

The test electrode shown in Figure 2.5b comprises a square metallic plate with sides that measure 250 mm and a square of dampened water-absorbent paper or cloth, from which surplus water has been removed, with sides that measure approximately 270 mm. The paper/cloth is placed between the metal plate and the surface being tested. During measurement a force of approximately 750 N (75 kg in weight) for floors or 250 N for walls is applied on the plate.

▼ **Figure 2.5b**
Test electrode 2

Fig 13B

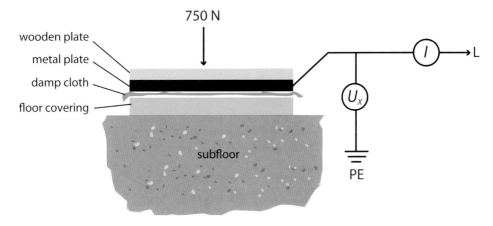

Expected results

The floors and walls are considered to be non-conducting where the measured resistances are at least 50 kΩ (where the system voltage to earth does not exceed 500 V).

612.5.2 A further test is specified in BS 7671 for extraneous-conductive-parts that are within the location but to which insulation has been applied during construction. In these cases a 'flash' insulation tester is required which, after the standard 500 V insulation test, applies a 2000 V a.c. rms test and measures the leakage current (which should be no more than 1 mA).

612.6 ### 2.7.12 Polarity testing

The polarity of all circuits must be verified before connection to the supply, with either an ohmmeter or the continuity range of an insulation and continuity tester. A typical test on a lighting circuit is shown in Figure 2.6.

Alternatively, polarity can be verified by visually checking core colours at terminations, thus verifying the installer's connections. Whatever method is used, polarity checks are required at all points on a circuit.

Instrument: Use a low-resistance ohmmeter for these test – see section 4.3.

It is necessary to check that all fuses and single-pole control and protective devices are connected in the line conductor. The centre contact of screw-type lampholders must be connected to the line conductor (except E14 and E27 to BS EN 60238).

Note: The continuity test (see 2.7.5) and ring continuity test (see 2.7.6) may confirm polarity.

CITY OF LIVERPOOL COLLEGE
VAUXHALL ROAD
LIVERPOOL
L3 6BN

▼ **Figure 2.6**
Polarity test on a
lighting circuit

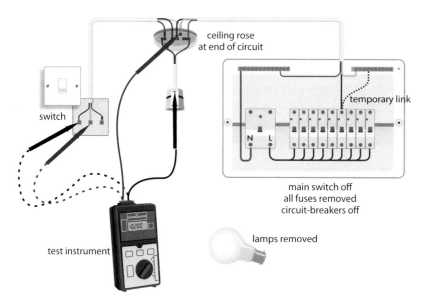

Note: the test may be carried out either at lighting points or switches

**REMEMBER TO REMOVE THE TEMPORARY SHORTING LINK WHEN TESTING
IS COMPLETE.**

612.7 **2.7.13 Earth electrode resistance testing**

542.1
542.2
Three methods of measuring the resistance of an earth electrode are described in this
section. Test method E1 uses a dedicated earth electrode tester (fall of potential,
three- or four-terminal type), test method E2 uses a dedicated earth electrode tester
(stakeless or probe type) and test method E3 uses an earth fault loop impedance
tester.

Test method E1: Measurement using dedicated earth electrode tester (fall of potential, three- or four-terminal type)

It is essential to ensure that the installation is securely isolated from the supply. It
is also necessary to disconnect the earthing conductor from the earth electrode.
**Caution: If this is the only earth electrode this may leave the installation
unprotected against earth faults and complete isolation of the installation
must be made.** This disconnection will ensure that the test current only passes
through the earth electrode and not through any parallel paths. The installation must
remain isolated from the supply until all testing has been completed and the earth
electrode connection reinstated.

Ideally, the test should be carried out when the ground conditions are least favourable,
such as during dry weather.

The test requires the use of two temporary test spikes (electrodes), and is carried out
in the following manner:

Connection to the earth electrode is made using terminals C1 and P1 of a four-terminal
earth tester. To exclude the resistance of the test leads from the resistance reading,
individual leads should be taken from these terminals and connected separately to
the electrode. Where the test lead resistance is insignificant, the two terminals may be
short-circuited at the test instrument and connection made with a single test lead, the
same being true if using a three-terminal tester. Connection to the temporary spikes is
made as shown in Figure 2.7.

▼ **Figure 2.7**
Typical earth electrode
test using a three- or
four-terminal tester

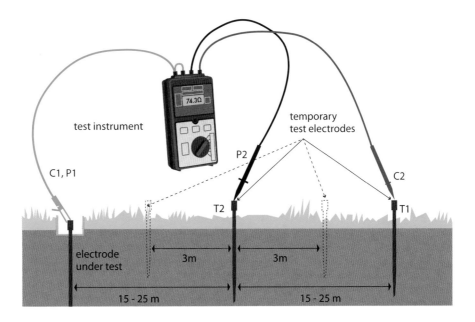

The distance between the test spikes is important. If they are too close together their resistance areas will overlap. In general, reliable results may be expected if the distance between the electrode under test and the current spike, C2, is at least ten times the maximum dimension of the electrode system, e.g. 30 m for a 3 m long rod electrode.

Three readings are taken:

▶ firstly, with the potential spike, T2, inserted midway between the electrode and the current spike, T1
▶ secondly, with T2 moved to a position 10 per cent of the overall electrode-to-current spike distance back towards the electrode
▶ last, with T2 moved to a position 10 per cent of the overall distance towards the current spike, from its initial position between the electrode and T1.

By comparing the three readings, a percentage deviation can be determined. This is calculated by taking the average of the three readings, finding the maximum deviation of the readings from this average in ohms, and expressing this as a percentage of the average.

The accuracy of the measurement using this technique is typically 1.2 times the percentage deviation of the readings. It is difficult to achieve an accuracy of measurement better than 2 per cent, and inadvisable to accept readings that differ by more than 5 per cent. In this event, to improve the accuracy of the measurement the test must be repeated with a larger separation between the electrode and the current spike.

The test instrument output may be a.c. or reversed d.c. to overcome electrolytic effects. Because these instruments employ phase-sensitive detectors, the errors associated with stray currents are eliminated.

The instrument should be capable of checking that the resistance of the temporary spikes used for testing is within the accuracy limits stated in the instrument specification. This may be achieved by an indicator provided on the instrument, or the instrument should have a sufficiently high upper range to enable a discrete test to be performed on the spikes.

Where the resistance of the temporary spikes is too high, measures to reduce the resistance will be necessary, such as driving the spikes deeper into the ground or watering with brine to improve the contact resistance. **In no circumstances should the latter technique be used to temporarily reduce the resistance of the earth electrode under test.**

ON COMPLETION OF THE TEST ENSURE THAT THE EARTHING CONDUCTOR IS RECONNECTED.

Test method E2: Measurement using dedicated stakeless earth electrode tester

612.7
542.1
542.2

While accurate, test method E1 is slow and requires disconnection of the earthing conductor.

Where there are a number of earth electrodes a 'stakeless' or clamp-based earth tester carries out electrode resistance measurements without disconnecting the earthing conductor.

Figure 2.8 shows the connection of such a tester in a two-transformer, four electrode set-up and the loop required for these testers (note, they cannot be used for measurement of a single electrode).

▼ **Figure 2.8**
Diagrammatic connection of a stakeless or probe-type tester

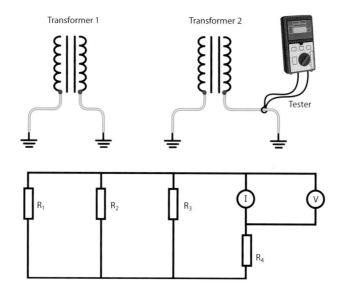

Test method E3: Measurement using an earth fault loop impedance tester

An earth electrode may be tested using an earth fault loop impedance tester. It is recognised that the results may not be as accurate as using a dedicated earth electrode tester.

SWITCH OFF SUPPLY BEFORE DISCONNECTING THE EARTHING CONDUCTOR. The earth fault loop impedance tester is connected between the line conductor at the source of the installation and the earth electrode, and a test performed. The impedance reading taken is treated as the electrode resistance.

ON COMPLETION OF THE TEST ENSURE THAT THE EARTHING CONDUCTOR IS RECONNECTED.

Results of earth electrode testing

For TN-S systems and generator supplies, electrode resistance values may not have been specified as electrodes often simply provide a local reference earth.

For TT systems, in the absence of the designer's specification, BS 7671 maximum values are as follows:

411.5.3 Regulation 411.5.3 requires:

$$R_A I_{\Delta n} \leq 50 \text{ V}$$

where:

R_A is the sum of the resistances of the earth electrode and the protective conductor(s) connecting it to the exposed-conductive-parts (in ohms)

$I_{\Delta n}$ is the rated residual operating current (in amperes).

Maximum values of Z_s, which may be substituted for R_A in the above equation, for residual current devices are given in Table 2.7.

▼ **Table 2.7** Maximum values of earth fault loop impedance (Z_s) for non-delayed RCDs to BS EN 61008-1 and BS EN 61009-1 for U_0 of 230 V

Table 41.5

RCD rated residual operating current, $I_{\Delta n}$ (mA)	Maximum value of earth fault loop impedance, Z_s (Ω)
30	1667
100	500
300	167
500	100

Where a delayed RCD is used to provide fault protection the maximum value of earth fault loop impedance including the earth electrode resistance must be such that the requirements of 411.3 and 411.5 are met. This is likely to require a lower figure than given above.

The table indicates that the use of a suitably rated RCD will theoretically allow much higher values of R_A, and therefore of Z_s, than could be expected by using the circuit overcurrent devices for fault protection.

It is advised that earth electrode resistance values above 200 ohms may not be stable, as soil conditions change due to factors such as soil drying and freezing.

2.7.14 Protection by automatic disconnection of supply

The effectiveness of measures for fault protection by automatic disconnection of supply can be verified for installations within a TN system by:

▶ measurement of earth fault loop impedance (as described in 2.7.15 below)
▶ confirmation by visual inspection that overcurrent devices have suitable short-time or instantaneous tripping setting for circuit-breakers, or current rating (I_n) and type for fuses
▶ where RCDs are employed, testing to confirm that the disconnection times of Chapter 41 of BS 7671 can be met (see 2.7.15 and 2.7.18).

For installations within a TT system, effectiveness can be verified by:

▶ measurement of the resistance of the earthing arrangement of the exposed-conductive-parts of the equipment for the circuit in question
▶ confirmation by visual inspection that overcurrent devices have suitable short-time or instantaneous tripping setting for circuit-breakers, or current rating (I_n) and type for fuses
▶ where RCDs are employed, testing to confirm that the disconnection times of Chapter 41 of BS 7671 can be met (see 2.7.15 and 2.7.18).

612.9 ## 2.7.15 Earth fault loop impedance verification

Where limitation of earth fault loop impedance is part of a protective measure, then it is fundamental that the initial verification process includes verification of earth fault loop impedances.

The earth fault current loop comprises the following elements, starting at the point of fault on the line–earth loop:

▶ the circuit protective conductor
▶ the main earthing terminal and earthing conductor
▶ for TN systems, the metallic return path or, in the case of TT and IT systems, the earth return path
▶ the path through the earthed neutral point of the transformer
▶ the source line winding and
▶ the line conductor from the source to the point of fault.

612.8 There are two methods used for verifying total earth fault loop impedance for a circuit:
612.9

▶ measurement of total earth fault loop impedance (Z_s) using an instrument
▶ measurement of ($R_1 + R_2$) during continuity testing of a circuit (see 2.7.5 and 2.7.6) and addition to the measured earth fault loop impedance external to that circuit (Z_e).

Which method is used comes down to a matter of personal choice and both are described below.

Measurement of total earth fault loop impedance (Z_s) using an instrument

Measurement of Z_s is made on a live installation and for safety and practical reasons, neither the connection with earth nor bonding conductors are disconnected.

Instrument: Use an earth fault loop impedance tester for this test – section 4.5.

Measurement of ($R_1 + R_2$) during continuity testing of a circuit and addition to the earth fault loop impedance external to that circuit (Z_e)

This procedure is described in section 2.7.5 and, for ring circuits, section 2.7.6, and the ($R_1 + R_2$) value recorded for a particular circuit is added to the earth fault loop impedance at the origin of that circuit.

For a consumer unit at the origin of an installation, this is as follows:

$$Z_s = Z_e + (R_1 + R_2)$$

where:

Z_s is the total earth fault loop impedance in ohms

Z_e is the external earth fault loop impedance, in this case 'external' to the installation

$(R_1 + R_2)$ is the measured resistance of the line conductor and circuit protective conductor, measured during the continuity test method 1 or ring circuit continuity test step 3.

For larger installations with consumer units or distribution boards not at the origin, there can arise confusion over the term 'external earth loop impedance' (Z_e) and some prefer to write or note the earth fault loop impedance at the distribution board as Z_{db} as, strictly speaking, this value is not external to the installation. Thus, the formula is denoted:

$$Z_s = Z_{db} + (R_1 + R_2)$$

Residual current devices

The test (measuring) current of earth fault loop impedance testers may trip any RCD protecting the circuit. This will prevent a measurement being taken and may result in an unwanted disconnection of supply to the circuit under test.

Instrument manufacturers can supply loop testers that are less liable to trip RCDs by either limiting the test current (to less than 15 mA) or by d.c. biasing (this technique saturates the core of the RCD prior to applying the test).

Measurement of external earth fault loop impedance, Z_e

The external earth fault loop impedance, Z_e, is measured using an earth fault loop impedance tester at the origin of the installation. The impedance measurement is made between the line of the supply and the means of earthing *with the main switch open or*

542.4.2   *with all the circuits isolated.* The means of earthing must be disconnected from the installation earthed equipotential bonding for the duration of the test to remove parallel

610.1   paths. Care should be taken to avoid any shock hazard to the testing personnel and other persons on the site both whilst establishing contact, and performing the test.

ENSURE THAT THE EARTH CONNECTION HAS BEEN REPLACED BEFORE RECLOSING THE MAIN SWITCH.

See Figure 2.9 for test method connections.

Instrument: Use an earth fault loop impedance tester for this test – see section 4.5.

▼ **Figure 2.9**
Example test of Z_e at the origin of an installation

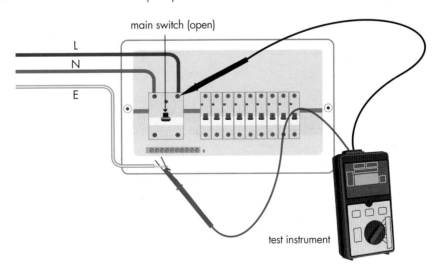

As described above, the measured Z_e can be used to add to circuit $(R_1 + R_2)$ values.

Determining external earth fault loop impedance, Z_e, by enquiry

The external earth fault loop impedance, Z_e, can be determined by enquiry to the electricity distributor. However, if this is relied upon, a test must be made to ensure that the distributor's earth terminal is actually connected with earth, using an earth fault loop impedance tester or a test lamp; it is usually easier simply to measure Z_e.

Verification of earth fault loop impedance test results

612.1   It is important to recognise that BS 7671 requires the inspector not only to test the installation but also to compare the results with relevant design criteria (or with criteria within BS 7671). This may seem obvious, but it is not uncommon for some inspectors to pass test information back to their office without making the necessary comparisons, possibly assuming that the office or someone else will check the results; the office may then assume that the inspector has checked the results against criteria, but no one has!

Values of Z_s should be compared with one of the following:

1   for standard thermoplastic (pvc) circuits, the values in Appendix A of this Guidance Note

Table 41.2   **2**   earth fault loop impedance figures provided by the designer. See also Appendix A,
Table 41.3   which provides information on how to correct measured results for ambient
Table 41.4   temperature as this may not have been done by the designer (the inspector will need to clarify this point)

      3   tabulated values in BS 7671, corrected for temperature. See Appendix A, which provides information on how to correct measured results for ambient temperature

      4   using a factor of 0.8, see Appendix A2.

Appendix A provides a formula for making temperature adjustments, together with a worked example.

612.11   ## 2.7.16 Prospective fault current, I_{pf}

Regulation 612.11 requires that the prospective fault current under both short-circuit and earth fault conditions be measured, calculated or determined by another method, at the origin and at other relevant points in the installation.

434.1   Regulation 612.11 introduces the requirements of Regulation 434.1 into the testing section, the designer being required to determine the prospective fault current, under both short-circuit and earth fault conditions, **at every relevant point of the installation**. This may be done by calculation, be ascertained by enquiry or be measured directly using an instrument. The expression 'every relevant point' means every point where a protective device is required to operate under fault conditions, and includes the origin of the installation.

The inspector must have knowledge of the design in this respect as, for example, if the switchgear at the origin of an installation is suitably rated for prospective fault current and switchgear of similar short-circuit rating is used downstream of that point, then no further checks are necessary as the magnitude of the prospective fault current will decrease.

434.5.1   Regulation 434.5.1 states that the breaking capacity rating of each protective device shall be not less than the prospective fault current at its point of installation. The term
Part 2   *prospective fault current* includes the prospective short-circuit current and the prospective earth fault current. It is the greater of these two prospective fault currents which should be determined and compared with the breaking capacity of the device.

With the power on, the **maximum value** of the prospective short-circuit current can be obtained by direct connection of the instrument between live conductors at the protective device at the origin or other relevant location within the installation. Both two-lead and three-lead instruments capable of determining prospective fault current are available and it is important that any instrument being used is set on the correct range and connected in accordance with the manufacturer's instructions for its use. Failure to do so could be dangerous, could result in damage to the instrument and may result in misleading readings being obtained.

Instrument: Use the prospective fault current range of a suitable earth fault loop impedance tester for this test – see section 4.5 (final paragraph).

With some instruments, the voltage between lines cannot be measured directly. Where this is the case, it can be assumed that for three-phase supplies, the maximum balanced prospective short-circuit level will be, as a rule of thumb, approximately twice the single-phase value. This figure errs on the side of safety.

Prospective earth fault current may be obtained with the same instrument. Again, care must be taken to ensure that the instrument is set correctly and connected as per the manufacturer's instructions for use.

The values obtained should be compared with the breaking capacity of the appropriate protective device. The breaking capacity of the protective device should be greater than the highest value of prospective fault current obtained using the instrument.

Whichever is the greater of the prospective short-circuit current and the prospective earth fault current obtained should be recorded on the Schedule of Test Results.

For a three-phase system, the prospective three-phase short-circuit current will always be larger than the single-phase line to neutral or earth fault currents.

Note on accuracy of earth fault loop impedance and prospective fault current testers (see also section 4.5)

Earth fault loop impedance testers become less accurate at smaller value readings. It should be noted that the standard instrument used for determining prospective fault current is effectively an earth loop impedance instrument. Accuracy can be variable on some instruments for readings below 1 ohm and accuracy is affected by the proximity of measurements to the supply transformer.

Instruments that use low currents (such as 15 mA) for determining earth fault loop impedance are more prone to error than using the higher current test on the same instrument.

Rated short-circuit breaking capacities of protective devices

The rated short-circuit capacities of fuses, circuit-breakers to BS EN 60898 and BS 3871 (now withdrawn) and RCBOs to BS EN 61009 are shown in Table 2.8. Note that BS 3871 identified the short-circuit capacity of circuit-breakers with an 'M' rating.

▼ **Table 2.8** Rated short-circuit capacities of protective devices

Device type	Device designation	Rated short-circuit capacity (kA)	
Semi-enclosed fuse to BS 3036 with category of duty	S1A S2A S4A	1 2 4	
General-purpose fuse to BS 88-2			
System E (bolted) type		80 kA at 400 V	
System G (clip in) type		50 kA at 230 V or 80 kA at 400 V	
Domestic fuse to BS 88-3			
Fuse system C type I type II		16 kA 31.5 kA	
BS 88-6		16.5 at 240 V 80 at 415 V	
Circuit-breakers to BS EN 60898* and RCBOs to BS EN 61009*		I_{cn} 1.5 3.0 6 10 15 20 25	I_{cs} (1.5) (3.0) (6.0) (7.5) (7.5) (10.0) (12.5)
BS 1361 Fuses†			
Domestic fuse to BS 1361 type 1 type 2		16.5 33.0	
Circuit-breakers to BS 3871 (replaced by BS EN 60898)	M1 M1.5 M3 M4.5 M6 M9	1 1.5 3 4.5 6 9	

* Two short-circuit capacity ratings are defined in BS EN 60898 and BS EN 61009:
     I_{cn} the rated short-circuit capacity (marked on the device)
     I_{cs} the service short-circuit capacity.
† BS 1361 will be/has been withdrawn from 1/9/2013.

The difference between the two short-circuit-ratings described above is the condition of the circuit-breaker after manufacturer's testing.

I_{cn} is the maximum fault current the device can interrupt safely, although it may no longer be usable.

I_{cs} is the maximum fault current the device can interrupt safely without loss of performance.

The I_{cn} value is marked on the device in a rectangle e.g. 6000 and for the majority of applications the prospective fault current at the terminals of the circuit-breaker should not exceed this value.

For domestic installations the prospective fault current is unlikely to exceed 6 kA, up to which value I_{cn} will equal I_{cs}.

For switchgear, the relevant fault current (short-circuit) rating of the switchgear (or assembly) should be equal to or exceed the maximum prospective fault current at the point of connection to the system. The terminology to define the short-circuit rating of an assembly is given in the BS EN 61439 series of standards as follows:

▶ rated short-time withstand current I_{cw}
▶ rated peak withstand current I_{pk}
▶ rated conditional short-circuit current I_{cc}.

Where a service cut-out containing a cartridge fuse to BS 88-3 (formerly BS 1361) supplies a consumer unit which complies with BS 5486-13 or BS EN 60439-3 Annex ZA, then the short-circuit capacity of the overcurrent protective devices within consumer units may be taken to be 16 kA.

Fault currents up to 16 kA

Except in London and some other major city centres, the maximum fault current for 230 V single-phase supplies up to 100 A is unlikely to exceed 16 kA.

The short-circuit capacity of overcurrent protective devices incorporated within consumer units may be taken to be 16 kA where:

▶ the consumer unit complies with BS 5486-13 or BS EN 60439-3
▶ the consumer unit is supplied through a type 2 fuse to BS 1361:1971 rated at no more than 100 A.

Recording the prospective fault current

Both the Electrical Installation Certificate and the Electrical Installation Condition Report contain a box headed Nature of Supply Parameters, which requires the prospective fault current at the origin to be recorded. The value to be recorded is the greater of either the short-circuit current (between live conductors) or the earth fault current (between line conductor(s) and the main earthing terminal). If it is considered necessary to record values at other relevant points, they can be recorded on the Schedule of Test Results. Where the protective devices used at the origin have the necessary rated breaking capacity, and devices with similar breaking capacity are used throughout the installation, it can be assumed that the Regulations are satisfied in this respect for all distribution boards.

2.7.17 Phase sequence testing

612.12   The 17th Edition introduced in Regulation 612.12 a requirement to verify that the phase sequence is maintained for multiphase circuits within an installation. In practice, this will be achieved by checking polarity and connections throughout the installation.

Optionally and occasionally, the inspector may wish to check phase sequence by using a phase rotation tester, either

▶ rotating disc type, or
▶ indicator lamp type.

Instruments containing both of the above forms of indication are also available.

Various types exist, a rotating disc, an electronic LCD equivalent or other means of indication. Generally, coloured or labelled leads are connected to the installation and if the phase sequence/rotation is correct the indication confirms this.

In the case of a rotating disc type instrument, the disc will be rotating either clockwise or anticlockwise.

With the indicator lamp type either the L1/L2/L3 (formerly R/Y/B) lamp or the L1/L3/L2 (formerly R/B/Y) lamp will be illuminated.

Both types of phase sequence indicator can also be used to verify phase sequence/ direction of rotation at the supply terminals to motors and to confirm the correct labelling/identification of plain conductors.

2.7.18 Operation and functional testing of RCDs

The operating times of RCDs are required to be tested in the following circumstances:

612.8.1 ▶ where they are relied on for disconnection for compliance with Chapter 41
612.10 ▶ where they are installed as additional protection as specified in Chapter 41.

Where RCDs are installed with circuit-breakers and the circuit has the characteristics to satisfy Chapter 41 without the RCD, then testing of the RCD is not essential unless it is specified for additional protection.

Operation of residual current devices

411.4.5 For each of the tests, readings should be taken on both positive and negative half-cycles and the longer operating time recorded.

Prior to these RCD tests it is essential, for safety reasons, that the earth loop impedance is tested to check the requirements have been met.

Instrument: Use an RCD tester for these tests, see section 4.7.

Test method

The test is made on the load side of the RCD between the line conductor of the protected circuit and the associated cpc. The load should be disconnected during the test. These tests can result in a potentially dangerous voltage on exposed-conductive-parts and extraneous-conductive-parts when the earth fault loop impedance approaches the maximum acceptable limits. Precautions must therefore be taken to prevent contact of persons or livestock with such parts.

The operating time should be no greater than those in Table 2.9, noting that all RCDs should first be tested at 50 per cent of rated current and must not operate/open with this test. This provides compliance with the appropriate product standard and will
411.4.5   provide compliance with Regulation 411.4.5.

▼ **Table 2.9** Operational tripping times for various RCDs

Device type	Non-time delayed	With time delay	Notes
	maximum operating time at 100% rated tripping current, $I_{\Delta n}$ (ms)	operating time at 100% rated tripping current, $I_{\Delta n}$ (ms)	
BS 4293	200	{(0.5 to 1.0) × time delay} + 200	
BS 61008	300	130 to 500	S type
BS 61009 (RCBO)	300	130 to 500	S type
BS 7288 (integral socket-outlet)	200	non-applicable	

Additional protection

415.1   Where an RCD with a rated residual operating current, $I_{\Delta n}$, not exceeding 30 mA is
612.10   used to provide additional protection in the event of failure of basic protection and/or the provision for fault protection or carelessness by users, the operating time of the device must not exceed 40 ms when subjected to a test current of 5 $I_{\Delta n}$. The maximum test time should not exceed 40 ms, unless the protective conductor potential rises by less than 50 V. (The instrument supplier will advise on compliance.)

Integral test device

612.13.1   An integral test device is incorporated in each RCD. This device enables the functioning of the mechanical parts of the RCD to be verified by pressing the button marked 'T' or 'Test'.

Operation of the integral test device does **not** provide a means of checking:

1   the continuity of the earthing conductor or the associated circuit protective conductors, or
2   any earth electrode or other means of earthing, or
3   any other part of the associated installation earthing, or
4   the sensitivity of the device.

The RCD test button will only operate the RCD if it is energized.

2.7.19 Other functional testing

612.13.2   All assemblies, including switchgear, controls and interlocks, should be functionally tested, that is operated to confirm that they work and are properly installed, mounted and adjusted.

2.7.20 Verification of voltage drop

612.14   Where it may be necessary to verify that voltage drop does not exceed the limits stated
Sect 525   in relevant product standards of installed equipment, BS 7671 provides two options to do so. Where no such limits are stated, voltage drop should be such that it does not impair the proper and safe functioning of installed equipment.

Voltage drop problems are quite rare but the inspector should be aware that long runs and/or high currents can sometimes cause voltage drop problems.

Measurement of voltage drop within an installation is not practical as this would mean measuring the instantaneous voltage at both the origin and at the point of interest simultaneously, together with the instantaneous load current.

Verification of voltage drop is not normally required during initial verification.

It is usually sufficient to check that voltage drop calculations have been undertaken.

Table 4Ab Appendix 4 of BS 7671 gives maximum values of voltage drop for lighting and for other uses, depending upon whether the installation is supplied directly from an LV distribution system or from a private LV supply.

It should be remembered that voltage drop may exceed the values stated in Appendix 4 in situations such as motor starting periods and where equipment has a high inrush current where such events remain within the limits specified in the relevant product standard or reasonable recommendation by a manufacturer.

2.7.21 Verification in medical locations

Sect 710 Medical locations were introduced into BS 7671:2008 by Amendment No. 1:2011.

710.61 The installation and testing of these installations is very much a specialist area and only the general requirements of BS 7671 are provided in this Guidance Note. Initial verification is carried out by an inspection and functional tests of the isolation IT system equipment including the insulation monitoring devices. Testing is required to measure the leakage current of the output circuit of medical IT isolating transformers and measurement of the resistance of the supplementary equipotential bonding.

2.7.22 Verification of electromagnetic disturbances

Sect 444 Inspectors should familiarise themselves with the new section on electromagnetic compatibility, Section 444 of BS 7671, which was introduced by Amendment No. 1: 2011.

It should be noted that compliance with EMC requirements in BS 7671 and in the EMC Regulations 2006 is something that is not verified by testing. The ethos of achieving compatibility is in design (with possibly some of the mitigating effects) and compliance is by way of storing the design criteria.

Section 444 specifies additional mitigating methods for EMC applied to the design and installation of cables and equipment. Many of these mitigating effects concern the routing of cables and their distance from other cables, as well as providing earth bonding.

Thus, verification of EMC and compliance with Section 444 is as follows:

► checking the EMC design with respect to cable routing, separation distances, enclosure etc.
► inspection of cable sheath and screen terminations and if considered necessary continuity checking of these items
► carrying out continuity checks of any additional mitigating bonding network provided (for example, a local mesh network).

It should be noted that there are no requirements for either installers or inspectors to carry out electric field or magnetic field strength measurements.

Periodic inspection and testing   3

3.1 Purpose of periodic inspection and testing

621.2 The purpose of periodic inspection and testing is to provide an engineering view on whether or not the installation is in a satisfactory condition where it can continue to be used safely.

A detailed visual examination of the installation is required, together with appropriate tests. The tests are mainly to confirm that the disconnection times stated in Chapter 41 are met.

The periodic inspection and testing is carried out, so far as is reasonably practicable, for:

1 the safety of persons and livestock against the effects of electric shock and burns
2 protection against damage to property by fire and heat arising from an installation defect
3 confirmation that the installation is not damaged or deteriorated so as to impair safety
4 the identification of installation defects and departures from the requirements of the Regulations that may give rise to danger.

622.2 For an installation under effective supervision in normal use, periodic inspection and testing may be replaced by an adequate regime of continuous monitoring and maintenance of the installation and all its constituent equipment by competent persons. It is important in such regimes that maintenance records, with references to inspection and testing, are recorded and stored. Such records should be available for scrutiny and need not be in the standard IET Electrical Installation Condition Report format.

3.2 Necessity for periodic inspection and testing

Periodic inspection and testing is necessary because all electrical installations deteriorate due to a number of factors such as damage, wear, tear, corrosion, excessive electrical loading, ageing and environmental influences. Consequently:

1 legislation requires that electrical installations are maintained in a safe condition, and this lends itself to periodic inspection and testing – see also Tables 3.1 and 3.2
2 licensing authorities, public bodies, insurance companies, mortgage lenders and others may require periodic inspection and testing of electrical installations, as is for example the case for houses in multiple occupation – see Tables 3.1 and 3.2
3 additionally, periodic inspection and testing should be considered in the following circumstances:
   ▶ to assess compliance with BS 7671

> ▶ on a change of occupancy of the premises
> ▶ on a change of use of the premises
> ▶ after additions or alterations to the original installation
> ▶ where there is a significant change (increase) in the electrical loading of the installation
> ▶ where there is reason to believe that damage may have been caused to the installation, as might be the case for example after flooding.

Reference to legislation and other documents is made below and it is vital that these requirements are ascertained before undertaking periodic inspection and testing.

3.3 Electricity at Work Regulations

Regulation 4(2) of the Electricity at Work Regulations 1989 requires that:

> As may be necessary to prevent danger, all systems shall be maintained so as to prevent, so far as is reasonably practicable, such danger.

The *Memorandum of guidance on the Electricity at Work Regulations 1989* (HSR25) published by the Health and Safety Executive advises that this regulation is concerned with the need for maintenance to ensure the safety of the system rather than being concerned with the activity of doing the maintenance in a safe manner, which is required by Regulation 4(3). The obligation to maintain a system arises if danger would otherwise result. There is no specific requirement to carry out a maintenance activity as such; what is required is that the system be kept in a safe condition. The frequency and nature of the maintenance must be such as to prevent danger so far as is reasonably practicable. Regular inspection of equipment including the electrical installation is an essential part of any preventive maintenance programme. This regular inspection may be carried out as required with or without dismantling and supplemented by testing.

There is no specific requirement to test the installation on every inspection. Where testing requires dismantling, the inspector should consider whether the risks associated with dismantling and reassembling are justified. Dismantling, and particularly disconnection of cables or components, introduces a risk of unsatisfactory reassembly.

3.4 Design

341.1 When carrying out the design of an installation and particularly when specifying the equipment, the designer should take into account the quality of the maintenance to be reasonably expected, including the frequency of routine checks and the period between subsequent inspections (supplemented as necessary by testing).

Information on the requirements for routine checks and inspections should be provided in accordance with Section 6 of the Health and Safety at Work etc. Act 1974 and as required by the Construction (Design and Management) Regulations 2007. Users of a premises should seek this information as the basis on which to make their own assessments. The Health and Safety Executive advise in their *Memorandum of guidance on the Electricity at Work Regulations 1989* (HSR25), that practical experience of an installation's use may indicate the need for an adjustment to the frequency of checks and inspections. This is a matter of judgement for the dutyholder. The Electrical Installation Certificate requires the designer's advice as to the intervals between inspections to be inserted on the certificate.

3.5 Routine checks

Electrical installations should not be left without any attention for the periods of years that are normally allowed between formal inspections. In domestic premises it is presumed that the occupier will soon notice any breakages or excessive wear and arrange for precautions to be taken and repairs to be carried out.

Commercial and industrial installations come under the Electricity at Work Regulations 1989 and formal arrangements are required for maintenance and interim routine checks (as well as periodic inspections); there should also be facilities to receive wear-and-tear reports from users of the premises.

The frequency and type of these routine checks will depend entirely upon the nature of the premises and should be set by the electrical dutyholder. Routine checks should include the items listed in Table 3.1. Table 3.2 (section 3.7) provides guidance on the frequency of routine checks, which may need to be increased as an installation ages.

▼ **Table 3.1** Routine checks

Activity	Check
Defects reports	All reported defects have been rectified

Inspection	Look for:
	    breakages
	    wear/deterioration
	    signs of overheating
	    missing parts (covers, screws)
	    loose fixings
	
	Confirm:
	    switchgear accessible (not obstructed)
	    doors of enclosures secure
	    adequate labelling in place

Operation	Operate:
	    switchgear (where reasonable)
	    equipment – switch on and off
	    including RCDs (using test button)

Note that routine checks need not be carried out by an electrically skilled person but should be done by somebody who is able to safely use the installation and recognise defects.

3.6 Required information

It is essential that the inspector knows the extent of the installation to be inspected and any criteria regarding the limit of the inspection. This should be recorded.

514.9 Enquiries should be made to the person responsible for the electrical installation with regard to the provision of diagrams, design criteria, type of electricity supply (and any alternative supply) and earthing arrangements.

Diagrams, charts or tables should be available to indicate the type and composition of circuits, identification of protective devices for shock protection, isolation and switching and a description of the method used for fault protection.

3.7 Frequency of periodic inspections

The time intervals between the recommended dates of periodic inspections require careful consideration. The date for the first periodic inspection and test is required to be considered and recommended by the installation designer. The date of each subsequent periodic inspection is required to be considered and recommended as part of carrying out a periodic inspection and test, by the person undertaking that particular inspection and test.

134.2.2

622.1 In determining the time interval for the date of the next recommended periodic inspection and test, the inspector is required to take into consideration the type of installation and equipment, its use and operation, any known maintenance and the external influences to which it is subjected.

Typical installation types with suggested initial frequencies between periodic inspection and testing are provided in Table 3.2. The competent person carrying out subsequent inspections may recommend that the interval between future inspections be increased or decreased as a result of the findings of their inspection. For example, the inspector may recommend an interval greater than that suggested by Table 3.2 in instances where an installation has not suffered from damage or deterioration which would detract from its overall suitability for continued use. Conversely, it would be appropriate to recommend that the next inspection is carried out sooner than is suggested in Table 3.2 where an installation has clearly not withstood the adverse effects of its environment and usage well and is not subject to adequate and appropriate maintenance.

In short, the inspector being a competent person should apply engineering judgement when deciding upon intervals between inspecting and testing an installation and may use the recommendations of Table 3.2 as a suitable starting point for such a decision.

In the case of domestic and commercial premises, a change in occupancy of the premises may necessitate additional inspection and testing.

621.2 The formal inspections should be carried out in accordance with Chapter 62 of BS 7671. This requires an inspection comprising a detailed examination of the installation, carried out without dismantling or with partial dismantling as required, together with the appropriate tests of Chapter 61.

▼ **Table 3.2**
Recommended initial frequencies of inspection of electrical installations

Type of installation	Routine check see section 3.5	Maximum period between inspections and testing (note 8)	Notes
General installation			
Domestic accommodation – general	–	Change of occupancy/10 years	
Domestic accommodation – rented houses and flats	1 year	Change of occupancy/5 years	1, 2, 10
Residential accommodation (Houses of Multiple Occupation) – halls of residence, nurses accommodation, etc.	1 year	Change of occupancy/5 years	1, 2, 10, 11
Commercial	1 year	Change of occupancy/5 years	1, 2, 3, 4
Educational establishments	6 months	5 years	1, 2, 6
Industrial	1 year	3 years	1, 2
Offices	1 year	5 years	1, 2
Shops	1 year	5 years	1, 2
Laboratories	1 year	5 years	1, 2
Hospitals and medical clinics			
Hospitals and medical clinics – general areas	1 year	5 years	1, 2
Hospitals and medical clinics – medical locations	6 months	1 year	9
Buildings open to the public			
Cinemas	1 year	1–3 years	2, 6
Church installations	1 year	5 years	2
Leisure complexes (excluding swimming pools)	1 year	3 years	1, 2, 6
Places of public entertainment	1 year	3 years	1, 2, 6
Restaurants and hotels	1 year	5 years	1, 2, 6
Theatres	1 year	3 years	2, 6, 7
Public houses	1 year	5 years	1, 2, 6
Village halls/community centres	1 year	5 years	1, 2
Special and specific installations (for medical locations see above)			
Agricultural and horticultural	1 year	3 years	1, 2
Caravans	1 year	3 years	7
Caravan parks	6 months	1 year	1, 2, 6
Highway power supplies	as convenient	6–8 years	
Marinas	4 months	1 year	1, 2
Fish farms	4 months	1 year	1, 2
Swimming pools	4 months	1 year	1, 2, 6
Emergency lighting	daily/monthly	3 years	2, 3, 4
Fire alarms	daily/weekly	1 year	2, 4, 5
Launderettes	monthly	1 year	1, 2, 6
Petrol filling stations	1 year	1 year	1, 2, 6
Construction site installations	3 months	3 months	1, 2

Notes:
1 Particular attention must be taken to comply with SI 2002 No. 2665 – Electricity Safety, Quality and Continuity Regulations 2002 (as amended).
2 Electricity at Work Regulations 1989, Regulation 4 and Memorandum of guidance (HSR 25) published by the HSE.
3 See BS 5266 – Part 1: 2005 *Code of practice for the emergency lighting of premises*.
4 Other intervals are recommended for testing operation of batteries and generators.

5   See BS 5839 – Part 1: 2002 + A2:2008 *Fire detection and alarm systems for buildings. Code of practice for system design, installation, commissioning and maintenance.*

6   Local Authority Conditions of Licence.

7   It is recommended that a caravan is inspected and tested every three years, reduced to every year if it is used frequently (see Regulation 721.514.1 and Fig 721 – Instructions for electricity supply).

8   The person carrying out subsequent inspections may recommend that the interval between future inspections be increased or decreased as a result of the findings of their inspection.

9   Medical locations shall have their isolating transformer equipment inspected and tested for functionality as well as alarms etc.; every third year the output leakage current of the IT isolating equipment shall be measured.

10 The Landlord & Tenant Act 1985 requires that properties under the Act have their services maintained. Periodic inspection and testing is the IET recognised method of demonstrating this.

11 The Management of Houses in Multiple Occupation Regulation (England and Wales).

3.8   Requirements for inspection and testing

3.8.1   Scope

The purpose of periodic inspection and testing is to provide an engineering view on whether or not the installation is in a satisfactory condition where it can continue to be used in a safe way.

The periodic inspection and test comprises a detailed examination of the installation together with appropriate tests. The inspection is carried out without taking apart or dismantling equipment as far as is possible. The tests made are mainly to confirm that the disconnection times stated in Chapter 41 are met, as well as highlighting other defects.

It is important that the competency of the person carrying out the periodic inspection and test is of the appropriate level having gained sufficient education, experience and knowledge to be fully conversant with the aspects required of carrying out such an important inspection. He or she will, for example, need to be able to inspect switchgear, determine the age of installation components and recognise signs of their deterioration. As well as having sufficient visual inspection skills they will also need to possess good testing skills and experience of older installations and knowledge of what the resulting data means in the context of the ongoing safety of the installation.

621.2   The requirement of BS 7671 for periodic inspection and testing is for a detailed inspection comprising an examination of the installation without dismantling, or with partial dismantling as required, together with the tests of Chapter 61 considered appropriate by the person carrying out the inspection and testing. The scope of the periodic inspection and testing must be decided by a competent person, taking into account the information contained in this section.

3.8.2   Process – prior to carrying out inspection and testing

Prior to carrying out the inspection, the inspector will need to meet with the client or the client's representative to outline the scope and nature of the work required and to highlight likely items that require isolation.

Consultation with the client or the client's representative prior to the periodic inspection and testing work being carried out is essential to determine the degree of disconnection which will be acceptable before planning the detailed inspection and testing. Also, the extent of previous maintenance, routine tests and documentation, including the original design and Electrical Installation Certificate, should be established.

For safety, it is necessary to carry out a visual inspection of the installation before testing or opening enclosures, removing covers, etc. So far as is reasonably practicable, the visual inspection must verify that the safety of persons, livestock and property is not endangered.

3.8.3 General procedure

Note: The following advice is not applicable to domestic or simple installations as the extent and method of inspection and testing is rudimentary in such installations in comparison with more complex installations.

Although there are various approaches to carrying out inspection and testing, one suggested method is to first obtain an overview of the installation, ideally from diagrams and charts as well as from a simple 'walk round' survey prior to starting the full inspection. This will enable the inspector to be able to plan the inspection and identify items that require isolation etc. Most importantly, this initial survey will enable the inspector to set sample sizes, see section 3.8.4.

621.1 Where diagrams, charts or tables are not available, a degree of exploratory work may be necessary so that inspection and testing can be carried out safely and effectively; this may include a survey to identify switchgear, controlgear and the circuits they control.

Indeed, for more involved installations without diagrams or charts the client should be advised that such diagrams require producing prior to the inspection and testing commencing. Alternatively, the inspection can commence in cases where the inspector feels that it is safe to proceed (this may be limited to visual inspection); the production of diagrams and charts can be called for on the Electrical Installation Condition Report.

Note should be made of any known changes in environmental conditions, building structure, and additions or alterations which have affected the suitability of the wiring for its present load and method of installation.

621.3 During the inspection, the opportunity should be taken to identify dangers which might arise during the testing. Any location and equipment for which safety precautions may be necessary should be noted and the appropriate steps taken.

A thorough inspection should be made of all electrical equipment which is not concealed, and should include the accessible internal condition of a sample of the equipment. The external condition should be noted and if damage is identified or if the degree of protection has been impaired, this should be recorded on the Schedule of Inspections appended to the Report. The inspection should include a check on the condition of electrical equipment and material, taking into account any available manufacturer's information, with regard to the following:

1 safety
2 age
3 damage
4 corrosion and external influence
5 overloading (signs of)
6 wear and tear and environment
7 suitability for continued use.

CITY OF LIVERPOOL COLLEGE
VAUXHALL ROAD
LIVERPOOL
L3 6BN

The assessment of condition should take account of known changes in conditions influencing and affecting electrical safety, for example plumbing or structural changes.

Where parts of an electrical installation are excluded from the scope of a Periodic Inspection and Test, they should be identified in the 'Extent and limitations' box of the Report.

Periodic tests should be made in such a way as to minimise disturbance of the installation and inconvenience to the user. Where it is necessary to disconnect part or the whole of an installation in order to carry out a test, the disconnection should be made at a time agreed with the user and for the minimum period needed to carry out the test. Where more than one test necessitates a disconnection, where possible they should be made during one disconnection period.

612.3.2
612.3.3
A careful check should be made of the type of equipment on site so that the necessary precautions can be taken, where conditions require, to disconnect or short-out electronic and other equipment which may be damaged by testing. Special care must be taken where control and protective devices contain electronic components.

3.8.4 Setting inspection and testing samples

The inspector must be familiar with setting both inspection and testing sample sizes as carrying out 100 per cent inspection or testing in many installations is unrealistic, uneconomical and unachievable. Information is provided in this section and the relevant sampling tables (for inspections see Table 3.3, for testing see Table 3.4).

As highlighted in 3.8.3, one recommended procedure is for the inspector to carry out an initial walk-round survey to establish initial sample sizes at various points throughout the installation. The detailed inspection is then started and the sample size is adjusted upwards if necessary, depending upon the results obtained. Samples should be selected that are representative of the whole installation. Parts of the installation that, in the inspector's experience are more likely to be problematic, should be prioritised.

The inspector will require all of his or her experience in setting sample sizes and should consider:

▶ approximate age and probable condition of the electrical installation
▶ type and usage of the installation or part thereof
▶ ambient environmental conditions
▶ the apparent effectiveness of ongoing maintenance, if any
▶ period of time elapsed since previous inspection/testing
▶ the size of the installation
▶ consultation with the installation owner
▶ the quality of records such as electrical installation certificates, minor works certificates, previous periodic inspection reports, maintenance records, site plans/drawings and data sheets relating to installed equipment.

It should be noted that the initial sample size is only based on a walk-round and consultation of records. Further, that what may at first appear to be good, e.g. the quality of maintenance, may turn out to be poor during the detailed inspection and testing.

▼ Figure 3.1
Suggested procedure for setting initial and adjusted sample sizes

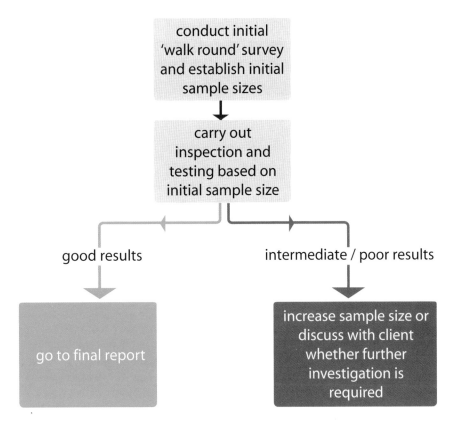

conduct initial 'walk round' survey and establish initial sample sizes

↓

carry out inspection and testing based on initial sample size

good results

intermediate / poor results

go to final report

increase sample size or discuss with client whether further investigation is required

Where the inspection or testing of a sample yields poor or unacceptable results this would suggest that similar problems may exist elsewhere in uninspected or untested items. The inspector will then need to either increase the sampling or refer back to the client; it may be that the inspector recommends that 100 per cent testing is carried out in that area.

The principle of this is indicated in Figure 3.1.

As an example, testing final circuits at a distribution board with a sample size of 10 per cent of the lighting circuits. More than one of these circuits tested with an unacceptably high earth fault loop impedance with a relatively low earth fault loop impedance at the incoming terminals to the distribution board itself. There were no apparent factors to suggest why the final circuit values were high. It would be remiss to complete the Electrical Installation Condition Report form by using just this 10 per cent sample and stating that improvements were required for these circuits. It would be far more appropriate to increase the sample size or recommend that all circuits at the distribution board were tested, based on the initial findings.

If relatively small sample sizes are chosen it is important that these are representative of the complete installation. Similarly, if a repeat periodic inspection is undertaken using a sampling system then a different sample, again representative of the complete installation, must be chosen. Therefore, previous periodic inspection and test records should be consulted prior to commencement of a sample inspection and test. Suggested sample sizes for visual inspections are provided in Table 3.3; suggested sample sizes for testing are discussed in section 3.10.1 and shown in Table 3.4.

▼ **Table 3.3** Range of samples for inspection

Item	Suggested minimum sample size (notes 1, 2)	Typical checks
Main switchgear external inspection	100%	Signs of damage, overheating or ageing.
Main switchgear internal sections and cable terminations	Ideally 100% but not less than 10% (note 2)	Signs of overheating, ageing, check tightness of cable connections
Main switchgear internal inspection of circuit-breaker connections and control sections	Ideally 100% but not less than 10%	Signs of overheating, ageing, check tightness of cable connections
Final circuit distribution boards	Ideally 100% but not less than 25% (note 4)	Signs of overheating, ageing, check of cable connections
Final circuit accessories	Between 10% to 100% (note 3)	Damage, signs of overheating
Earthing and protective bonding conductors	100%	Presence and tightness

Notes:
1 Where the inspection of a sample yields poor or unacceptable results this would suggest that similar problems may exist elsewhere in the uninspected items. The inspector will then need to either increase the sampling or refer back to the client; it may be that the inspector recommends that 100 per cent inspection is carried out in that area.
2 100 per cent where practicable.
3 Generally, it is less appropriate to apply small sample sizing to the inspection of socket-outlets compared to samples for lighting as it is more likely that user equipment will be hand-held and therefore of greater potential risk from electric shock.
4 Do not 'sample samples', resulting in a very low overall sampled installation. Samples must be representative. If it is decided to sample for example submain cables at 10 per cent then further sampling should not be applied to the final circuit distribution boards on these circuits.

3.9 Periodic inspection

3.9.1 Example checklist of items that require inspection

Appx 6   The following is a copy of the checklist in Appendix 6 of BS 7671:2008(2011), which lists items at various locations within an installation that may require inspection.

ELECTRICAL INTAKE EQUIPMENT

- ▶ Service cable
- ▶ Service cut-out/fuse
- ▶ Meter tails – Distributor
- ▶ Meter tails – Consumer
- ▶ Metering equipment
- ▶ Isolator

Where inadequacies in distributor's equipment are encountered, it is recommended that the person ordering the report informs the appropriate authority.

PRESENCE OF ADEQUATE ARRANGEMENTS FOR PARALLEL OR SWITCHED ALTERNATIVE SOURCES (551.6; 551.7)

[The inspector may feel it necessary to prove interlocking arrangements where provided.]

AUTOMATIC DISCONNECTION OF SUPPLY

▶ Main earthing/bonding arrangements (411.3; Chap 54)
   1 Presence of distributor's earthing arrangement (542.1.2.1; 542.1.2.2), or presence of installation earth electrode arrangement (542.1.2.3)
   2 Adequacy of earthing conductor size (542.3; 543.1.1)
   3 Main protective earthing conductor connections (542.3.2)
   4 Accessibility of earthing conductor connections (543.3.2)
   5 Adequacy of main protective bonding conductor sizes (544.1)
   6 Main protective bonding conductor connections (543.3.2; 544.1.2)
   7 Accessibility of all protective bonding connections (543.3.2)
   8 Provision of earthing/bonding labels at all appropriate locations (514.11)
▶ FELV

OTHER METHODS OF PROTECTION

(Where any of the methods listed below are employed details should be provided on separate sheets.)

▶ Non-conducting location (418.1)
▶ Earth-free local equipotential bonding (418.2)
▶ Electrical separation (Section 413; 418.3)
▶ Double insulation (Section 412)
▶ Reinforced insulation (Section 412)

DISTRIBUTION EQUIPMENT

▶ Adequacy of working space/accessibility to equipment (132.12; 513.1)
▶ Security of fixing (134.1.1)
▶ Condition of insulation of live parts (416.1)
▶ Adequacy/security of barriers (416.2)
▶ Condition of enclosure(s) in terms of IP rating etc (416.2)
▶ Condition of enclosure(s) in terms of fire rating etc (421.1.6; 526.5)
▶ Enclosure not damaged/deteriorated so as to impair safety (621.2(iii))
▶ Presence and effectiveness of obstacles (417.2)
▶ Placing out of reach (417.3)
▶ Presence of main switch(es), linked where required (537.1.2; 537.1.4)
▶ Operation of main switch(es) (functional check) (612.13.2)
▶ Manual operation of circuit-breakers and RCDs to prove disconnection (612.13.2)
▶ Confirmation that integral test button/switch causes RCD(s) to trip when operated (functional check) (612.13.1)
▶ RCD(s) provided for fault protection – includes RCBOs (414.4.9; 411.5.2; 531.2)
▶ RCD(s) provided for additional protection, where required - includes RCBOs (411.3.3; 415.1)
▶ Presence of RCD quarterly test notice at or near equipment, where required (514.12.2)
▶ Presence of diagrams, charts or schedules at or near equipment, where required (514.9.1)

- ▶ Presence of non-standard (mixed) cable colour warning notice at or near equipment, where required (514.14)
- ▶ Presence of alternative supply warning notice at or near equipment, where required (514.15)
- ▶ Presence of next inspection recommendation label (514.12.1)
- ▶ Presence of other required labelling (please specify) (Section 514)
- ▶ Examination of protective device(s) and base(s); correct type and rating (no signs of unacceptable thermal damage, arcing or overheating) (421.1.3)
- ▶ Single-pole protective devices in line conductor only (132.14.1; 530.3.2)
- ▶ Protection against mechanical damage where cables enter equipment (522.8.1; 522.8.11)
- ▶ Protection against electromagnetic effects where cables enter ferromagnetic enclosures (521.5.1)

DISTRIBUTION CIRCUITS

- ▶ Identification of conductors (514.3.1)
- ▶ Cables correctly supported throughout their run (522.8.5)
- ▶ Condition of insulation of live parts (416.1)
- ▶ Non-sheathed cables protected by enclosure in conduit, ducting or trunking (521.10.1)
- ▶ Suitability of containment systems for continued use (including flexible conduit) (Section 522)
- ▶ Cables correctly terminated in enclosures (Section 526)
- ▶ Examination of cables for signs of unacceptable thermal or mechanical damage/deterioration (421.1; 522.6)
- ▶ Adequacy of cables for current-carrying capacity with regard for the type and nature of installation (Section 523)
- ▶ Adequacy of protective devices: type and rated current for fault protection (411.3)
- ▶ Presence and adequacy of circuit protective conductors (411.3.1.1; 543.1)
- ▶ Coordination between conductors and overload protective devices (433.1; 533.2.1)
- ▶ Cable installation methods/practices with regard to the type and nature of installation and external influences (Section 522)
- ▶ Where exposed to direct sunlight, cable of a suitable type (522.11.1)
- ▶ Cables concealed under floors, above ceilings, in walls/partitions less than 50 mm from a surface, and in partitions containing metal parts
  - **1** installed in prescribed zones (see Section D. *Extent and limitations*) (522.6.101) or
  - **2** incorporating earthed armour or sheath, or run within earthed wiring system, or otherwise protected against mechanical damage by nails, screws and the like (see Section D. *Extent and limitations*) (522.6.101; 522.6.103)
- ▶ Provision of fire barriers, sealing arrangements and protection against thermal effects (Section 527)
- ▶ Band II cables segregated/separated from Band I cables (528.1)
- ▶ Cables segregated/separated from non-electrical services (528.3)
- ▶ Condition of circuit accessories (621.2(iii))
- ▶ Suitability of circuit accessories for external influences (512.2)
- ▶ Single-pole devices for switching in line conductor only (132.14.1; 530.3.2)
- ▶ Adequacy of connections, including cpc's, within accessories and to fixed and stationary equipment – identify/record numbers and locations of items inspected (Section 526)

- Presence, operation and correct location of appropriate devices for isolation and switching (537.2)
- General condition of wiring systems (621.2(ii))
- Temperature rating of cable insulation (522.1.1; Table 52.1)

FINAL CIRCUITS

- Identification of conductors (514.3.1)
- Cables correctly supported throughout their run (522.8.5)
- Condition of insulation of live parts (416.1)
- Non-sheathed cables protected by enclosure in conduit, ducting or trunking (521.10.1)
- Suitability of containment systems for continued use (including flexible conduit) (Section 522)
- Adequacy of cables for current-carrying capacity with regard for the type and nature of installation (Section 523)
- Adequacy of protective devices: type and rated current for fault protection (411.3)
- Presence and adequacy of circuit protective conductors (411.3.1.1; 543.1)
- Co-ordination between conductors and overload protective devices (433.1; 533.2.1)
- Wiring system(s) appropriate for the type and nature of the installation and external influences (Section 522)
- Cables concealed under floors, above ceilings, in walls/partitions less than 50 mm from a surface, and in partitions containing metal parts
  1 installed in prescribed zones (see Section D. *Extent and limitations*) (522.6.101)
  2 incorporating earthed armour or sheath, or run within earthed wiring system, or otherwise protected against mechanical damage from by nails, screws and the like (see Section D. *Extent and limitations*) (522.6.101; 522.6.103) or
  3 *for an installation not under the supervision of skilled or instructed persons, provided with additional protection by a 30 mA RCD (522.6.102; 522.6.103)
- Provision of additional protection by 30 mA RCD
  1 *for circuits used to supply mobile equipment not exceeding 32 A rating for use outdoors in all cases (411.3.3)
  2 *for all socket-outlets of rating 20 A or less provided for use by ordinary persons unless exempt (411.3.3)
- Provision of fire barriers, sealing arrangements and protection against thermal effects (Section 527)
- Band II cables segregated/separated from Band I cables (528.1)
- Cables segregated/separated from non-electrical services (528.3)
- Termination of cables at enclosures – identify/record numbers and locations of items inspected (Section 526)
  1 Connections under no undue strain (526.6)
  2 No basic insulation of a conductor visible outside enclosure (526.8)
  3 Connections of live conductors adequately enclosed (526.5)
  4 Adequately connected at point of entry to enclosure (glands, bushes etc.) (522.8.5)
- Condition of accessories including socket-outlets, switches and joint boxes (621.2 (iii))

▶ Suitability of accessories for external influences (512.2)

*Note: Older installations designed prior to BS 7671:2008 may not have been provided with RCDs for additional protection

ISOLATION AND SWITCHING

▶ Isolators (537.2)
  1 Presence and condition of appropriate devices (537.2.2)
  2 Acceptable location – state if local or remote from equipment in question (537.2.1.5)
  3 Capable of being secured in the OFF position (537.2.1.2)
  4 Correct operation verified (612.13.2)
  5 Clearly identified by position and /or durable marking (537.2.2.6)
  6 Warning label posted in situations where live parts cannot be isolated by the operation of a single device (514.11.1; 537.2.1.3)
▶ Switching off for mechanical maintenance (537.3)
  1 Presence and condition of appropriate devices (537.3.1.1)
  2 Acceptable location – state if local or remote from equipment in question (537.3.2.4)
  3 Capable of being secured in the OFF position (537.3.2.3)
  4 Correct operation verified (612.13.2)
  5 Clearly identified by position and /or durable marking (537.3.2.4)
▶ Emergency switching/stopping (537.4)
  1 Presence and condition of appropriate devices (537.4.1.1)
  2 Readily accessible for operation where danger might occur (537.4.2.5)
  3 Correct operation verified (537.4.2.6)
  4 Clearly identified by position and /or durable marking (537.4.2.7)
▶ Functional switching (537.5)
  1 Presence and condition of appropriate devices (537.5.1.1)
  2 Correct operation verified (537.5.1.3; 537.5.2.2)

CURRENT–USING EQUIPMENT (PERMANENTLY CONNECTED)

▶ Condition of equipment in terms of IP rating etc (416.2)
▶ Equipment does not constitute a fire hazard (Section 421)
▶ Enclosure not damaged/deteriorated so as to impair safety (621.2(iii))
▶ Suitability for the environment and external influences (512.2)
▶ Security of fixing (134.1.1)
▶ Cable entry holes in ceiling above luminaires, sized or sealed so as to restrict the spread of fire: List number and location of luminaires inspected (separate page)
▶ Recessed luminaires (downlighters)
  1 Correct type of lamps fitted
  2 Installed to minimise build-up of heat by use of 'fire rated' fittings, insulation displacement box or similar (421.1.1)
  3 No signs of overheating to surrounding building fabric (559.5.1)
  4 No signs of overheating to conductors/terminations (526.1)

PART 7 SPECIAL INSTALLATIONS OR LOCATIONS

▶ If any special installations or locations are present, list the particular inspections applied.

3.10 Periodic testing

3.10.1 General

621.2 The periodic testing is supplementary to the inspection of the installation, see 3.8.1.

The same range and level of testing as for initial testing is not necessarily required, or indeed possible. Installations that have been previously tested and for which there are comprehensive records of test results may not need the same degree of testing as installations for which no such records exist.

Periodic testing may cause danger if the correct procedures are not applied. Persons carrying out periodic testing must be competent in the use of the instruments employed and have adequate knowledge and experience of the type of installation, see 3.8.1.

The inspector will need to set a sample size for testing. Notes on the principle of this are explained in 3.8.4, which should be studied together with the suggested tests of Table 3.4.

Where a sample test indicates results significantly different to those previously recorded, further investigation is necessary. Also, if during the course of testing a sample, significant errors were found that would suggest that the same problems may exist in untested items, then the inspector has to take appropriate action. This action needs to be either increasing the sampling or referring back to the client; it may be that the inspector recommends that 100 per cent testing is carried out in that area. This principle was shown in Figure 3.1.

3.10.2 Tests to be made

621.2 The tests considered appropriate by the person carrying out the inspection should be carried out in accordance with the recommendations in Table 3.4 and considering sections 3.8.1 to 3.8.4 of this Guidance Note.

See section 2.7 of this Guidance Note for test methods, noting that alternative methods may be used.

▼ **Table 3.4** Testing to be carried out where practicable on existing installations (see notes 1 and 2)

Test	Recommendations
Protective conductors continuity	Accessible exposed-conductive-parts of current-using equipment and accessories (notes 4 and 5)
Bonding conductors continuity	▶ Main bonding conductors to extraneous-conductive-parts ▶ Supplementary bonding conductors
Ring circuit continuity	Where there are records of previous tests, this test may not be necessary unless there may have been changes made to the ring final circuit
Polarity	At the following positions: ▶ origin of the installation ▶ distribution boards ▶ accessible socket-outlets ▶ extremity of radial circuits
Earth fault loop impedance	At the following positions: ▶ origin of the installation ▶ distribution boards ▶ accessible socket-outlets ▶ extremity of radial circuits *(contd)*

Test	Recommendations
Insulation resistance	If tests are to be made: ▶ between live conductors and Earth at main and final distribution boards (note 6)
Earth electrode resistance	If tests are to be made: ▶ test each earth rod or group of rods separately, with the test links removed, and with the installation isolated from the supply source
Functional tests RCDs	Tests as required by Regulation 612.13.1, followed by operation of the integral test button
Functional tests of circuit-breakers, isolators and switching devices	Manual operation to confirm that the devices disconnect the supply

Notes:

1   The person carrying out the testing should decide which of the above tests are appropriate by using their experience and knowledge of the installation being inspected and tested and by consulting any available records, see 3.8.4 of this Guidance Note.
2   Where sampling is applied, the percentage used is at the discretion of the inspector, see 3.8.4 of this Guidance Note (a percentage of less than 10 per cent is inadvisable).
3   The tests need not be carried out in the order shown in the table.
4   The earth fault loop impedance test may be used to confirm the continuity of protective conductors at socket-outlets and at accessible exposed-conductive-parts of current-using equipment and accessories.
5   Generally, accessibility may be considered to be within 3 m from the floor or from where a person can stand.
6   Where the circuit includes surge protective devices (SPDs) or other electronic devices which require a connection to earth for functional purposes, these devices will require disconnecting to avoid influencing the test result and to avoid damaging them.

3.10.3 Additional notes on periodic testing

This section provides some notes on the practicalities of carrying out the periodic tests, particularly within an installation where only partial isolation is possible.

a   Continuity of protective conductors and equipotential bonding conductors and earth fault loop impedance testing

If an electrical installation is isolated from the supply, it is permissible to disconnect protective and equipotential bonding conductors from the main earthing terminal in order to verify their continuity.

Where an electrical installation cannot be isolated from the supply, the protective and equipotential bonding conductors should **not** be disconnected as, under fault conditions, the exposed- and extraneous-conductive-parts could be raised to a dangerous level above earth potential. Also, measurement of earth fault loop impedance at various parts of the installation, is, for practical reasons, carried out with the protective earthing and bonding conductors connected.

A convenient way to carry out the above periodic tests in a large installation could be to use the wandering lead method to test continuity (see 2.7.5, test method 2) and to directly measure earth fault loop impedance at the same time.

Motor circuits

Loop impedance tests on motor circuits can only be carried out on the supply side of isolated motor controlgear. A continuity test between the circuit protective conductor and the motor is then necessary.

b Insulation resistance

Insulation resistance tests should be made on electrically isolated circuits with any electronic equipment which might be damaged by application of the test voltage disconnected, or only a measurement to protective earth made with the live conductors connected together.

For most installations the most practical test is an insulation test between live conductors (connected together) and earth; in practice time does not usually allow for a line to neutral test.

Check that information/warnings are given at the distribution board of circuits or equipment likely to be damaged by testing. Any diagram, chart or table should also include this warning.

The results of insulation testing should be compared with previous results where possible. Table 2.2 of this Guide (Table 61 of BS 7671) requires a minimum insulation resistance of 1 megohm, but strictly speaking this value applies to initial verification. It can, however, be used as a guide for periodic testing.

Where equipment is disconnected for these tests and the equipment has exposed-conductive-parts required by the Regulations to be connected to protective conductors, the insulation resistance between the exposed-conductive-parts and all live parts of the equipment should be measured separately and should comply with the requirements of the appropriate British Standard for the equipment.

There is a range of possible outcomes when carrying out insulation testing. Tests are typically made between all live conductors connected together and earth at a test voltage of 500 V d.c.

The inspector will need to measure the values of insulation resistance for a given distribution board and then take a view based on his/her engineering judgement as to whether the results obtained are acceptable. It should be noted that distribution boards with large numbers of final circuits will generally give a lower insulation resistance value than distribution boards with fewer final circuits.

c Polarity

It should be established whether there have been any additions or alterations to the installation since its last inspection. If there have been no additions or alterations then this test is a good example of where sampling can be applied.

For example, the following sampling could be used:

> 10 per cent of all single-pole and multi-pole control devices and of any centre-contact lampholders, together with 100 per cent of socket-outlets. If any incorrect polarity is found then a full test should be made in that part of the installation supplied by the particular distribution board concerned, and the sample testing increased for the remainder of the installation (say to 25 per cent); if additional cases of incorrect polarity are found in the 25 per cent sample, a full test of the complete installation should be made, see Figure 3.1.

d Operation of overcurrent circuit-breakers

612.13.2 Where protection against overcurrent is provided by circuit-breakers, the manual operating mechanism of each circuit-breaker should be operated to verify that the device opens and closes satisfactorily.

It is not normally necessary or practicable to test the operation of the automatic tripping mechanism of circuit-breakers. Any such test would need to be made at a current substantially exceeding the minimum tripping current in order to achieve operation within a reasonable time. For circuit-breakers to BS EN 60898 a test current of not less than two and a half times the nominal rated tripping current of the device is needed for operation within 1 minute, and much larger test currents are necessary to verify operation of the mechanism for instantaneous tripping.

For circuit-breakers of the sealed type, designed not to be maintained, if there is doubt about the integrity of the automatic mechanism it will normally be more convenient to replace the device than to make further tests. Such doubt may arise from visual inspection, if the device appears to have suffered damage or undue deterioration, or where there is evidence that the device may have failed to operate satisfactorily in service.

Circuit-breakers with the facility for injection testing may be so tested and, if appropriate, relay settings confirmed.

e Operation of devices for isolation and switching

Where means are provided in accordance with the requirements of the Regulations for isolation and switching, the devices should be operated to verify their effectiveness and checked to ensure adequate and correct labelling.

Easy access to such devices must be maintained, and effective operation must not be impaired by any material placed near the device. Access and operation areas may be required to be marked to ensure they are kept clear.

For isolating devices in which the position of the contacts or other means of isolation is externally visible, visual inspection of operation is sufficient and no testing is required.

The operation of every safety switching device should be checked by operating the device in the manner normally intended to confirm that it performs its function correctly in accordance with the requirements of BS 7671.

Where it is a requirement that the device interrupts all the supply conductors, the use of a test lamp or instrument connected between each line and the neutral on the load side of the switching device may be necessary. Reliance should not be placed on a simple observation that the equipment controlled has ceased to operate.

Where switching devices are provided with detachable or lockable handles in accordance with the Regulations, a check should be made to verify that the handles or keys are not interchangeable with any others available within the premises.

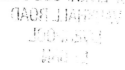

612.13.2 Where any form of interlocking is provided, e.g. between a main circuit-breaker and an outgoing switch or isolation device, the integrity of the interlocking must be verified; this may be beyond the scope of the inspector and something that is referred to a manufacturer or specialist.

Where switching devices are provided for isolation or for mechanical maintenance switching, the integrity of the means provided to prevent any equipment from being unintentionally or inadvertently energised or reactivated must be verified.

3.11 Electrical Installation Condition Report
(formerly called Periodic Inspection Report)

634.1 Amendment No 1 to BS 7671:2008 made extensive revisions to the reporting of periodic inspection and testing. The new report form is entitled Electrical Installation Condition Report and a model, together with model forms i.e. schedules of inspections and test results, are provided in Appendix 6 of BS 7671. Typical completed forms are given in Chapter 5 of this Guidance Note.

The full Electrical Installation Condition Report documentation comprises the following:

Electrical Installation Condition Report

and

schedule of inspections (one or more)

and

schedule of test results (one or more)

On completion of periodic inspection and testing, this Electrical Installation Condition Report and its accompanying schedules of inspections and schedules of test results must be given to the client or person who ordered the inspection.

634.2
Part 2 A most important point to remember is that any damage, deterioration, defects, dangerous conditions and non-compliance with BS 7671 that may give rise to danger (*danger* being a risk or injury to persons or livestock) must be recorded on the report.

There are new classification codes C1 to C3 for danger and non-compliances with BS 7671, these being explained in Table 3.5.

Each separate item entered in the Section K Observations box of the Report should be coded C1, C2 or C3 as appropriate.

CITY OF LIVERPOOL COLLEGE
VAUXHALL ROAD
LIVERPOOL
L3 6BN

▼ **Table 3.5** Classification of danger and non-compliances (for use during periodic inspection and testing)

Classification of danger or non-compliance	Description	Notes and guidance
C1	Danger present. Risk of injury. Immediate remedial action required	To be cited in situations which cannot be left. It is suggested that these are rectified or possibly, isolation may be recommended or necessary. Examples include accessible bare live parts, badly damaged equipment with risk of access to live parts, incorrect polarity, arcing found in switchgear.
C2	Potentially dangerous – urgent remedial action required	To be cited in situations that, whilst urgent, do not require immediate remedial action. Examples include a non-earthed installation (this requires a further fault to manifest injury), fundamentally undersized cables, earth fault loop impedance values greater than BS 7671 requirements, a 'borrowed' neutral, equipment with inappropriately selected IP (this may be C1 if severe), insulation readings under 1 MΩ, connections not terminated within appropriate enclosures.
C3	Improvement required	To be cited where C1 or C2 do not apply. Examples include the absence of most warning notices, absence of the required diagrams and charts, no or incorrect marking of conductors at terminations, absence of an RCD specified for additional protection (where the circuit otherwise tests as normal).

3.12 Periodic inspection of installations to an earlier edition of BS 7671 or the IEE Wiring Regulations

People often ask what standard should be applied when carrying out the periodic inspection of an installation constructed in accordance with an earlier edition of BS 7671, or an even earlier edition of the IEE Wiring Regulations or to an unknown standard.

In this situation the inspection should be carried out against the current edition of BS 7671. However, it is likely that there will be items that do not comply with the current edition of BS 7671 but this does not necessarily mean that the installation is unsafe. If the inspector feels that an item requires improvement, it should be given code C3 on the Electrical Installation Condition Report. If the finding does not require improvement it does not need to be recorded as an observation.

Reference is made to existing installations both in the second paragraph of the Introduction to BS 7671:2008 and in the Note by the Health and Safety Executive which follows the Preface to BS 7671.

Test instruments and equipment

4

4.1 Instrument standard

BS EN 61010 *Safety requirements for electrical equipment for measurement, control, and laboratory use* is the basic safety standard for electrical test instruments.

The basic instrument standard is BS EN 61557 *Electrical safety in low voltage distribution systems up to 1000 V a.c. and 1500 V d.c. Equipment for testing, measuring or monitoring of protective measures.* This standard includes performance requirements and requires compliance with BS EN 61010.

In Section 1.1, reference was made to the use of test leads which conform to HSE Guidance Note GS 38. The safety measures and procedures set out in GS 38 should be observed for all instruments, leads, probes and accessories. It should be noted that some test instrument manufacturers advise that their instruments be used in conjunction with fused test leads and probes. Other manufacturers advise the use of non-fused leads and probes when the instrument has in-built electrical protection, but it should be noted that such electrical protection does not extend to the probes and leads.

4.2 Instrument accuracy

A basic measurement accuracy of 5 per cent is usually adequate for these test instruments. In the case of analogue instruments, a basic accuracy of 2 per cent of full-scale deflection will provide the required accuracy measurement over a useful proportion of the scale.

It should not be assumed that the accuracy of the reading taken in normal field use will be as good as the basic accuracy. The 'operating accuracy' is always worse than the basic accuracy, and additional errors derive from three sources:

1 *Instrument errors*: basic instrument accuracy applies only in ideal conditions; the actual reading accuracy will also be affected by the operator's ability, battery condition, generator cranking speed, ambient temperature and orientation of the instrument.

2 *Loss of calibration*: instruments should be regularly recalibrated using standards traceable to National Standards, or have their accuracy cross-checked using known references such as comparing readings to those obtained from other instruments, or by the use of a proprietary instrument 'check box' having clearly defined characteristics. Where using other instruments as a calibration check the references used for checking instruments need to have a greater accuracy than required of the instrument being checked.

In all cases the type and frequency of recalibration or checking required should be as specified by the instrument manufacturer, taking into account ambient environmental and usage factors as appropriate. For example, if an instrument is left in storage at a constant temperature in a dry environment for long periods and is used infrequently, the user may be able to extend the recalibration interval. However, if an instrument is roughly handled and is regularly transported and stored in vehicles, and hence is subjected to fluctuations in temperature and humidity caused by changes in time of day/night and time of year, then more frequent confirmation of accuracy would be appropriate.

Instruments should also be subjected to regular checks so that errors caused by deterioration of leads, probes, connectors etc. do not result in inaccurate readings being recorded when, for example, schedules of test results are compiled.

It is essential that the accuracy of instruments is confirmed after any incidences of mechanical or electrical mishandling.

3 *Field errors*: the instrument reading accuracy will also be affected by external influences as a result of working in the field environment. These influences can take many forms, and some sources of inaccuracy are described in the appropriate sections.

BS EN 61557 requires a maximum operating error of ±30 per cent of reading over the stated measurement range.

To achieve satisfactory in-service performance, it is essential to be fully informed about the test equipment, how it is to be used, and the accuracy to be expected.

The accuracy and repeatability of the earth loop impedance measurements at low values is limited by inherent instrument errors and the occurrence of mains transients (and spikes), interference and also close vicinity to distribution transformers. Traditionally, in analogue instruments this corresponds to making measurements very near to the zero point on the scale where inaccuracies and non-repeatability are not at all evident with this type of electromechanical movement.

Traceability to National Standards can be assured by using a calibration laboratory accredited by a National Accreditation Body. In the UK this is the United Kingdom Accreditation Service (UKAS). A list of accredited laboratories can be found at www.ukas.com or a search for sources of calibration by instrument can be made at www.ukas.org/calibration

4.3 Low-resistance ohmmeters

612.2.1 The instrument used for low-resistance tests may be either a specialised low-resistance ohmmeter or the continuity range of an insulation and continuity tester. The test current may be d.c. or a.c. It is recommended that it be derived from a source with no-load voltage between 4 V and 24 V, and a short-circuit current not less than 200 mA.

The measuring range should cover the span 0.2 Ω to 2 Ω, with a resolution of at least 0.01 Ω for digital instruments.

Instruments to BS EN 61557-4 will meet the above requirements.

Field effects contributing to in-service errors are contact resistance, test lead resistance, a.c. interference and thermocouple effects in mixed metal systems.

Whilst contact resistance cannot be eliminated with two-terminal testers, and can introduce errors, the effects of lead resistance can be eliminated by measuring this prior to a test, and subtracting the resistance from the final value. Interference from an external a.c. source (interference pick-up) cannot be eliminated, although it may be indicated by vibration of the pointer of an analogue instrument. Thermocouple effects can be eliminated by reversing the test probes and averaging the resistance readings taken in each direction.

4.4 Insulation resistance testers

The instrument used should be capable of developing the test voltage required across the load.

The test voltage required is:

Table 61 **1** 250 V d.c. for SELV and PELV circuits
2 500 V d.c. for all circuits rated up to and including 500 V, but excluding extra-low voltage circuits mentioned above
3 1000 V d.c. for circuits rated above 500 V up to 1000 V.

Instruments conforming to BS EN 61557-2 will fulfil all the above instrument requirements.

The factors affecting in-service reading accuracy include 50 Hz currents induced into cables under test, and capacitance in the test object. These errors cannot be eliminated by test procedures. Capacitance may be as high as 5 µF, and the instrument should have an automatic discharge facility capable of safely discharging such a capacitance. Following an insulation resistance test, the instrument should be left connected until the capacitance within the installation has fully discharged.

4.5 Earth fault loop impedance testers

These instruments operate by circulating a current from the line conductor into the protective earth. This will raise the potential of the protective earth system.

To minimise electric shock hazard from the potential of the protective conductor, the test duration should be within safe limits. This means that the instrument should cut off the test current after 40 ms or a time determined by the safety limits derived from the information contained within DD IEC/TS 60479-1, if the voltage rise of the protective conductor exceeds 50 V during the test.

Instrument accuracy decreases as scale reading reduces. Aspects affecting in-service reading accuracy include transient variations of mains voltage during the test period, mains interference, test lead resistance and errors in impedance measurement as a result of the test method. To allow for the effect of transient voltages the test should be repeated at least once. The other effects cannot be eliminated by test procedures.

For circuits rated up to 50 A, a line–earth loop tester with a resolution of 0.01 Ω should be adequate.

Instruments conforming to BS EN 61557-3 will fulfil the above requirements.

These instruments may also offer additional facilities for deriving prospective fault current. The basic measuring principle is generally the same as for earth fault loop impedance testers. The current is calculated by dividing the earth fault loop impedance value into the mains voltage. Instrument accuracy is determined by the same factors as for loop testers. In this case, instrument accuracy decreases as scale reading increases, because the loop value is divided into the mains voltage. It is important to note these aspects, and the manufacturer's documentation should be referred to.

The instruments are not as accurate at low resolutions (low loop impedances and correspondingly high prospective fault currents) as they are at higher values and some manufacturers state that results should be treated with caution below (say) a half or one or two ohms. Whilst results should be treated with caution, reliance is placed on these instruments and they have generally proved satisfactory in these tests (calculations may be necessary to cross-check test results).

4.6   Earth electrode resistance testers

There are three general methods referred to in Chapter 2:

▶   method E1, using a dedicated earth electrode tester (fall of potential, three- or four-terminal type)
▶   method E2, using a dedicated earth electrode tester (stakeless or probe type)
▶   method E3, using an earth fault loop impedance tester.

The most accurate of these is method E1.

As mentioned in Chapter 2, method E2, the stakeless or probe tester, cannot be used to measure the resistance of a single earth electrode.

4.7   RCD testers

The test instrument should be capable of applying the full range of test current to an in-service accuracy as given in BS EN 61557-6. This in-service reading accuracy will include the effects of voltage variations around the nominal voltage of the tester.

To check RCD operation and to minimise danger during the test, the test current should be applied for no longer than 2 s.

Instruments conforming to BS EN 61557-6 will fulfil the above requirements.

4.8   Phase rotation instruments

BS EN 61557-7 gives the requirements for measuring equipment for testing the phase sequence in three-phase distribution systems whether indication is given by mechanical, visual and/or audible means.

BS EN 61557-7 includes requirements that:

▶   indication shall be unambiguous between 85 per cent and 110 per cent of the nominal system voltage or within the range of the nominal voltage and between 95 per cent and 105 per cent of the nominal system frequency
▶   the measuring equipment should be suitable for continuous operation
▶   the measuring equipment should be so designed that when either one or two measuring leads are connected to earth and the remaining measuring lead(s) remain connected to their corresponding line conductors, the resulting total current to earth should not exceed 3.5 mA rms
▶   the measuring equipment should not be damaged nor should the user be exposed to danger in situations where the measuring equipment is connected to 120 per cent of the rated system voltage or to 120 per cent of its rated maximum voltage range

▶ portable measuring equipment should be provided with permanently attached leads or with a plug device with live parts not accessible, whether plugged or unplugged.

4.9 Thermographic equipment

Although thermographic surveying is not recognised by BS 7671 as a test instrument, such equipment can be invaluable in assisting electrical inspections and as such some notes are included on this type of equipment.

Important note: It is recommended that persons refer to the requirements of the Electricity at Work Regulations 1989 and the guidance given in the HSE *Memorandum of guidance on the Electricity at Work Regulations 1989* (HSR25) prior to undertaking any work activity which places themselves or those under their control in close proximity to live parts.

It is relatively easy to make arrangements to disconnect small installations such as domestic premises from the supply to facilitate periodic inspection and testing.

However, as the size and complexity of an installation increases, isolation of the supply becomes increasingly difficult. This is particularly true where continuity of supply has health implications, as may be the case in hospitals and similar premises, or financial implications as would be the case in banks, share-dealing and commodities markets and the like. Nevertheless, it remains necessary to confirm the continuing suitability of such installations for use. Therefore they must still be subjected to planned and preventive maintenance or regular periodic assessment of their condition.

It may well be possible to carry out a thorough visual inspection of such installations without subjecting the inspector or others in the building to any danger, and such an inspection may identify many common defects caused by use/abuse. Furthermore, experience of such installations may provide a valuable insight into commonly occurring cases of wear and tear.

Some defects cannot be discovered by visual inspection alone. For example, incorrectly tightened connections can result in a high resistance joint which can then cause a high temperature to occur locally to the connection. If left uncorrected over time further deterioration of the connection may well occur, leading to a continuing increase in temperature which may subsequently present a risk of fire. This fire risk will be significantly increased in installations where a build-up of dust or other flammable materials can occur in close proximity to the source of heat. It should also be remembered that increased heat at terminations can result in accelerated deterioration of the insulation locally. Heating effects symptomatic of a fault or other problem within an electrical installation can also occur as a result of cyclical-load operations, use of conductors of inadequate current-carrying capacity, incorrect load balancing and more mechanically related issues such as incorrect alignment of motor drive couplings and overtightened belt-drives.

A number of manufacturers offer infrared imaging equipment which can be used to identify such 'hot spots'. Infrared thermography works on the principle that all materials emit electromagnetic radiation in the infrared region which can be detected by a thermal imaging camera.

The amount of radiated energy detected can be presented in a readily usable form, typically being shown as differences in colour that vary with the temperature being

4

▼ Figure 4.1
A colour/
temperature
correlation indicator

▼ Figure 4.2
Bolted connections
at a busbar as seen
by the eye

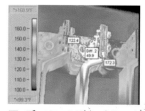

▼ Figure 4.3
Bolted connections
at a busbar viewed
using thermal
imaging

▼ Figure 4.4
Thermal image of
contactor showing
termination on right
is too hot

detected. Figure 4.1 shows a colour/temperature correlation indicator such as those that may accompany images. Such scales will aid the person ordering the inspection or responsible for maintenance activities in their interpretation of the thermal images.

In terms of visual inspection, the busbar connections in Figure 4.2 appear to be satisfactory.

However, if the same connections are viewed using a thermal imaging camera (Figure 4.3), it is evident that the connections to the centre are running significantly hotter than those to either side.

This higher temperature may indicate a loose connection or connections, but in this case is probably due to the centre bar carrying a significantly higher current than those to each side of it. As such, the person carrying out the inspection could suggest their client looks into improving the load balancing of this part of the installation.

In a further example, Figure 4.4 shows a contactor and the thermal image has highlighted a potential loose termination. After this was tightened Figure 4.5 shows the result, with all three terminations now operating at a much more uniform temperature.

Whilst the remedial work was being carried out it would be sensible to inspect the insulation of the conductors in the terminations to confirm that the insulation remained effective and had not suffered significant damage.

The requirements of the Electricity at Work Regulations 1989 must be taken into account when considering the use of thermographic surveying equipment as its use may necessitate the temporary removal or bypassing of measures that provide basic protection (as defined in BS 7671), such as opening doors to electrical panels and the removal of barriers and covers. The requirements of Regulation 14 (Work on or near live conductors), which is reproduced below, are particularly pertinent:

No person shall be engaged in any work activity on or so near any live conductor (other than one suitably covered with insulating material so as to prevent danger) that danger may arise unless –

(a) it is unreasonable in all the circumstances for it to be dead; and
(b) it is reasonable in all the circumstances for him to be at work on or near it while it is live; and
(c) suitable precautions (including where necessary the provision of suitable protective equipment) are taken to prevent injury.

The *Memorandum of guidance on the Electricity at Work Regulations 1989* (HSR25) recognises that it may be necessary, in some circumstances, for conductors to remain live during testing or diagnostic work. However, such work in close proximity to live conductors may only be carried out if it can be done safely and if all precautions required to allow it to be done so are put in place. Additionally, the work may only be performed by persons who are suitably competent with regard to the type and nature of the work activity being performed, as required by Regulation 16 (Persons to be competent to prevent danger or injury). HSR25 also makes clear that although live testing

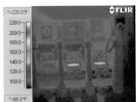

▼ Figure 4.5

Thermal image of contactor after loose termination has been tightened

may be justifiable it does not follow that there will necessarily be justification for subsequent repair work to be carried out live.

Persons carrying out thermographic surveying should:

▶ have sufficient competence to prevent danger and injury

▶ understand the system being worked on, the hazards that may arise as a result of the work and the precautions that are required to prevent danger

▶ be able to identify those parts of equipment being inspected which are, or are capable of being, live when the supply to the equipment is switched on

▶ implement all precautions required to prevent injury that have been identified as part of the risk assessment for the work

▶ maintain the maximum possible distance from the live or potentially live parts described above at all times

▶ maintain effective control of the area in which the equipment being inspected is situated

▶ ensure that all protective measures which may have been affected by their actions when carrying out the inspection work are fully reinstated. All guards and barriers must be replaced and panel doors, lids and covers must be closed and secured properly after the inspection is completed.

As demonstrated above, thermographic inspection can be an effective method of identifying defects that would not be located by a more conventional visual inspection. However, such thermal surveying should not be seen as a substitute for periodic inspection and testing, but rather as an additional tool that can be used by the inspector. Thermographic surveys can be a highly effective means of targeting preventive maintenance to where it is most required. Defects identified may be factored into the planned maintenance programme for the installation, or where necessary may justify the remedial work being performed without delay.

CITY OF LIVERPOOL COLLEGE
VAUXHALL ROAD
LIVERPOOL
L3 6BN

4

Forms

This chapter provides guidance on completing the necessary forms and certificates associated with inspection and testing. Sample certificates, completed with typical entries, are provided together with sample test result sheets and sample inspection sheets, again completed with typical results.

The chapter also contains some notes on completion of the forms, although helpful information on this will also be gained by reading earlier chapters of this Guidance Note.

5.1 Initial verification (inspection and testing) forms

Following an initial verification, or an addition or alteration to an existing installation, an Electrical Installation Certificate is required to be completed and issued together with inspection schedules and test result schedules.

There are two options for the Electrical Installation Certificate provided, Form 1 or Form 2 as follows:

▶ Form 1 – Short form of Electrical Installation Certificate (to be used when one person is responsible for the design, construction, inspection and testing of an installation)

or

▶ Form 2 – Electrical Installation Certificate (three signatory version from Appendix 6 of BS 7671)

Whichever Electrical Installation Certificate is used, appropriate numbers of the following forms are required to accompany the Certificate:

▶ Form 3 – Schedule of Inspections, and
▶ Form 4 – Generic Schedule of Test Results.

For completeness, two samples of typical completed Form 4s (schedule of test results) are included, one being for a single-phase installation and the other for a three-phase installation.

5.2 Minor works

The complete set of forms for initial inspection and testing may not be appropriate for minor works. When an addition to an electrical installation does not extend to the installation of a new circuit, the minor works form may be used. This form is intended for such work as the addition of a socket-outlet or lighting point to an existing circuit, or for repair or modification.

The Minor Electrical Installation Works Certificate (Form 5) is included and is taken from Appendix 6 of BS 7671. Notes on completion and guidance for recipients are provided with the form.

5.3   Periodic inspection and testing

Periodic inspection and testing is reported using the Electrical Installation Condition Report, see Form 6. This is used together with the appropriate number of generic schedule of test result sheets (Form 4). For periodic inspections the schedule of inspections for initial verification, i.e. Form 3 should not be used but the dedicated model suggested in BS 7671 as shown in Form 7 can be used. It is suggested that this is suitable for inspections of domestic installations and installations up to 100 amperes. For larger and more complex installations the inspector will need to formulate his/her own inspection schedules.

5.4   Model forms for certification and reporting

The introduction to Appendix 6 of BS 7671:2008 'Model forms for certification and reporting' is reproduced on this page.

Introduction

(i)   The Electrical Installation Certificate required by Part 6 should be made out and signed or otherwise authenticated by a competent person or persons in respect of the design, construction, inspection and testing of the work.

(ii)   The Minor Works Certificate required by Part 6 should be made out and signed or otherwise authenticated by a competent person in respect of the design, construction, inspection and testing of the minor work.

(iii)   The Electrical Installation Condition Report required by Part 6 should be made out and signed or otherwise authenticated by a competent person in respect of the inspection and testing of an installation.

(iv)   Competent persons will, as appropriate to their function under (i) (ii) and (iii) above, have a sound knowledge and experience relevant to the nature of the work undertaken and to the technical standards set down in these Regulations, be fully versed in the inspection and testing procedures contained in these Regulations and employ adequate testing equipment.

(v)   Electrical Installation Certificates will indicate the responsibility for design, construction, inspection and testing, whether in relation to new work or further work on an existing installation.

Where design, construction, inspection and testing are the responsibility of one person a Certificate with a single-signature declaration in the form shown below may replace the multiple signatures section of the model form.

FOR DESIGN, CONSTRUCTION, INSPECTION & TESTING

I being the person responsible for the Design, Construction, Inspection & Testing of the electrical installation (as indicated by my signature below), particulars of which are described above, having exercised reasonable skill and care when carrying out the Design, Construction, Inspection & Testing, hereby CERTIFY that the said work for which I have been responsible is to the best of my knowledge and belief in accordance with BS 7671:2008, amended to ....(date) except for the departures, if any, detailed as follows.

(vi) A Minor Works Certificate will indicate the responsibility for design, construction, inspection and testing of the work described on the certificate.

(vii) An Electrical Installation Condition Report will indicate the responsibility for the inspection and testing of an existing installation within the extent and limitations specified on the report.

(viii) Schedules of inspection and schedules of test results as required by Part 6 should be issued with the associated Electrical Installation Certificate or Electrical Installation Condition Report.

(ix) When making out and signing a form on behalf of a company or other business entity, individuals should state for whom they are acting.

(x) Additional forms may be required as clarification, if needed by ordinary persons, or in expansion, for larger or more complex installations.

CITY OF LIVERPOOL COLLEGE
VAUXHALL ROAD
LIVERPOOL
L3 6BN

Form 1   Form No: 505513..../1

ELECTRICAL INSTALLATION CERTIFICATE
(REQUIREMENTS FOR ELECTRICAL INSTALLATIONS - BS 7671 [IET WIRING REGULATIONS])

DETAILS OF THE CLIENT Mr T Brown
32 South St
Anytown, Surrey .... Post Code: TO1 1ZZ

INSTALLATION ADDRESS The Coffee Bean
31 Station Road
Anytown, Surrey .... Post Code: TO3 2YF

DESCRIPTION AND EXTENT OF THE INSTALLATION Tick boxes as appropriate

Description of installation:	New installation ☑
Re-wire of ground floor, on change of use.	
	Addition to an existing installation ☐
Extent of installation covered by this Certificate:	
Complete electrical re-wire of refurbished premises, on change of use from offices to cafe/snack bar.	Alteration to an existing installation ☐
(Use continuation sheet if necessary) see continuation sheet No:	

FOR DESIGN, CONSTRUCTION, INSPECTION & TESTING
I being the person responsible for the design, construction, inspection & testing of the electrical installation (as indicated by my signature below), particulars of which are described above, having exercised reasonable skill and care when carrying out the design, construction, inspection & testing hereby CERTIFY that the said work for which I have been responsible is to the best of my knowledge and belief in accordance with BS 7671:2008, amended to 2011...... (date) except for the departures, if any, detailed as follows:

Details of departures from BS 7671 (Regulations 120.3 and 133.5):
None

The extent of liability of the signatory is limited to the work described above as the subject of this Certificate.

Signature: *W Hastings* Date: 21-Jan-2011 Name (IN BLOCK LETTERS): W HASTINGS ....

Company Hastings Electrical ....
Address: 21 The Arches
.... Anytown, Surrey .... Postcode: TO2 9YY. Tel No: 01022 999999 ....

NEXT INSPECTION
I recommend that this installation is further inspected and tested after an interval of not more than ...5....... years/months.

SUPPLY CHARACTERISTICS AND EARTHING ARRANGEMENTS Tick boxes and enter details, as appropriate

Earthing arrangements	Number and Type of Live Conductors	Nature of Supply Parameters	Supply Protective Device Characteristics
TN-C ☐ TN-S ☐ TN-C-S ☑ TT ☐ IT ☐	a.c. ☑ d.c. ☐ 1-phase, 2-wire ☑ 2-wire ☐ 2-phase, 3-wire ☐ 3-wire ☐	Nominal voltage, $U/U_0{}^{(1)}$ ....230. V Nominal frequency, $f^{(1)}$ ....50. Hz	Type BS 1361 Fuse
	3-phase, 3-wire ☐ other ☐	Prospective fault current, $I_{pf}{}^{(2)}$ ..9.0. kA	Rated current.....100....A
Other sources ☐ of supply (to be detailed on attached schedules)	3-phase, 4-wire ☐ Confirmation of supply polarity ☑	External loop impedance, $Z_e{}^{(2)}$0.28. Ω *(Note: (1) by enquiry, (2) by enquiry or by measurement)*	

Page 1 of .4.

Form 1   Form No: <u>505513</u> ..../1

PARTICULARS OF INSTALLATION REFERRED TO IN THE CERTIFICATE *Tick boxes and enter details, as appropriate*

Means of Earthing | **Maximum Demand**

Distributor's facility ☑   Maximum demand (load) ....80..kVA / Amps *Delete as appropriate*

                                     Details of Installation Earth Electrode (*where applicable*)

Installation | Type | Location | Electrode resistance to Earth
earth electrode ☐ | (e.g. rod(s), tape etc) | |
                     | ....N/A.......... | ....N/A........ | ....N/A.... Ω

Main Protective Conductors

Earthing conductor: material Copper............ csa ....16....mm²   Continuity and connection verified ☑

Main protective bonding
conductors   material Copper............ csa ....10....mm²   Continuity and connection verified ☑

To incoming water and/or gas service ☑   To other elements: N/A..

Main Switch or Circuit-breaker

BS, Type and No. of poles BS EN 60947-3 (2-pole)   Current rating ....100..A   Voltage rating ....230....V

Location Services cupboard adjacent rear exit   Fuse rating or setting.........N/A..A

Rated residual operating current $I_{\Delta n}$ = ...N/A.. mA, and operating time of ......N/A.. ms (at $I_{\Delta n}$) *(applicable only where an RCD is suitable and is used as a main circuit-breaker)*

COMMENTS ON EXISTING INSTALLATION (in the case of an addition or alteration see Section 633):
Not Applicable..
....
....
....

SCHEDULES
The attached Schedules are part of this document and this Certificate is valid only when they are attached to it.
........1.... Schedules of Inspections and .....1...... Schedules of Test Results are attached.
(Enter quantities of schedules attached).

Page 2 of .4.

Form 2   Form No: 555513..../2

ELECTRICAL INSTALLATION CERTIFICATE
(REQUIREMENTS FOR ELECTRICAL INSTALLATIONS - BS 7671 [IET WIRING REGULATIONS])

DETAILS OF THE CLIENT	Mr D Roberts	
	23 Acacia Avenue	Post Code: SL0 0LT
	Sometown, Berks.	

INSTALLATION ADDRESS Unit 3 The Quadrant
Sometown Business Park .... Post Code: SL1 0ZZ
Sometown Berks.

DESCRIPTION AND EXTENT OF THE INSTALLATION Tick boxes as appropriate

Description of installation: Complete electrical installation to new 2 floor office building	New installation ☑
Extent of installation covered by this Certificate: Complete electrical installation, comprising main switchboard, sub-main and distribution boards. To include all power and lighting circuits but excluding car-park lighting, which is supplied from adjacent building.	Addition to an existing installation ☐
(Use continuation sheet if necessary) see continuation sheet No:	Alteration to an existing installation ☐

FOR DESIGN
I/We being the person(s) responsible for the design of the electrical installation (as indicated by my/our signatures below), particulars of which are described above, having exercised reasonable skill and care when carrying out the design hereby CERTIFY that the design work for which I/we have been responsible is to the best of my/our knowledge and belief in accordance with BS 7671:2008, amended to 2011..... (date) except for the departures, if any, detailed as follows:

Details of departures from BS 7671 (Regulations 120.3 and 133.5):
None

The extent of liability of the signatory or the signatories is limited to the work described above as the subject of this Certificate.

For the DESIGN of the installation: **(Where there is mutual responsibility for the design)

Signature: *D Jones*   Date: 20-Jan-2012   Name (IN BLOCK LETTERS): D JONES .... Designer No 1

Signature: *N/A*   Date: ....   Name (IN BLOCK LETTERS): .... Designer No 2**

FOR CONSTRUCTION
I/We being the person(s) responsible for the construction of the electrical installation (as indicated by my/our signatures below), particulars of which are described above, having exercised reasonable skill and care when carrying out the construction hereby CERTIFY that the construction work for which I/we have been responsible is to the best of my/our knowledge and belief in accordance with BS 7671:2008, amended to 2011....(date) except for the departures, if any, detailed as follows:

Details of departures from BS 7671 (Regulations 120.3 and 133.5):
None

The extent of liability of the signatory is limited to the work described above as the subject of this Certificate.

For CONSTRUCTION of the installation:

Signature: *T Smith*   Date: 20-Jan-2012   Name (IN BLOCK LETTERS): T SMITH .... Constructor

FOR INSPECTION & TESTING
I/We being the person(s) responsible for the inspection & testing of the electrical installation (as indicated by my/our signatures below), particulars of which are described above, having exercised reasonable skill and care when carrying out the inspection & testing hereby CERTIFY that the work for which I/we have been responsible is to the best of my/our knowledge and belief in accordance with BS 7671:2008, amended to 2011.....(date) except for the departures, if any, detailed as follows:

Details of departures from BS 7671 (Regulations 120.3 and 133.5):
None

The extent of liability of the signatory is limited to the work described above as the subject of this Certificate.

For INSPECTION AND TESTING of the installation:

Signature: *G Wilson*   Date: 20-Jan-2012   Name (IN BLOCK LETTERS): G WILSON .... Inspector

NEXT INSPECTION
I/We the designer(s), recommend that this installation is further inspected and tested after an interval of not more than 5........ years/months.

Page 1 of .4.

Form 2   Form No: 555513 ../2

PARTICULARS OF SIGNATORIES TO THE ELECTRICAL INSTALLATION CERTIFICATE

Designer (No 1)
Name: D Jones   Company: The Electrical Design Partnership
Address: 23 High Street
Sometown, Berks.   Postcode: SL 10 0YY   Tel No: 01000 999999

Designer (No 2)
(if applicable)
Name: ....   Company: ....
Address: ....
....   Postcode: ....   Tel No: ....

Constructor
Name: T Smith   Company: T Smith Electrical Installations
Address: Unit 8a Sometown Ind Estate
Sometown, Berks.   Postcode: SL3 0XX   Tel No: 01000 888888

Inspector
Name: G WILSON   Company: Wilson and Sons
Address: 11 Crabtree Row
Sometown, Berks.   Postcode: SL2 0WW   Tel No: 01000 777777

SUPPLY CHARACTERISTICS AND EARTHING ARRANGEMENTS Tick boxes and enter details, as appropriate

Earthing arrangements	Number and Type of Live Conductors	Nature of Supply Parameters	Supply Protective Device Characteristics
TN-C ☐	a.c. ☑ d.c. ☐	Nominal voltage, $U/U_0^{(1)}$ 400/230 V	Type: BS 1361 Fuse
TN-S ☐	1-phase, 2-wire ☐ 2-wire ☐	Nominal frequency, $f^{(1)}$ 50 Hz	
TN-C-S ☑	2-phase, 3-wire ☐ 3-wire ☐	Prospective fault current, $I_{pf}^{(2)}$ 0.77 kA	Rated current 100 A
TT ☐	3-phase, 3-wire ☐ other ☐	External loop impedance, $Z_e^{(2)}$ 0.30 Ω	
IT ☐	3-phase, 4-wire ☑		
Other sources ☐ of supply (to be detailed on attached schedules)	Confirmation of supply polarity ☑	(Note: (1) by enquiry, (2) by enquiry or by measurement)	

PARTICULARS OF INSTALLATION REFERRED TO IN THE CERTIFICATE Tick boxes and enter details, as appropriate

Means of Earthing | **Maximum Demand**

Distributor's facility ☑   Maximum demand (load) .... 50 kVA / ~~Amps~~ Delete as appropriate

Details of Installation Earth Electrode (where applicable)

Installation earth electrode ☐	Type (e.g. rod(s), tape etc)	Location	Electrode resistance to Earth
	N/A	N/A	N/A Ω

Main Protective Conductors

Earthing conductor: material Copper   csa 16 mm²   Continuity and connection verified ☑

Main protective bonding conductors   material Copper   csa 16 mm²   Continuity and connection verified ☑

To incoming water and/or gas service ☑   To other elements: N/A

Main Switch or Circuit-breaker

BS, Type and No. of poles BS EN 60947-3 (4-pole)   Current rating 100 A   Voltage rating 400 V

Location Main switchroom adjacent Reception area   Fuse rating or setting N/A A

Rated residual operating current $I_{\Delta n}$ = N/A mA, and operating time of N/A ms (at $I_{\Delta n}$) (applicable only where an RCD is suitable and is used as a main circuit-breaker)

COMMENTS ON EXISTING INSTALLATION (in the case of an addition or alteration see Section 633):
Not Applicable..

SCHEDULES
The attached Schedules are part of this document and this Certificate is valid only when they are attached to it.
......1.... Schedules of Inspections and .....1...... Schedules of Test Results are attached.
(Enter quantities of schedules attached).

Page 2 of 4

ELECTRICAL INSTALLATION CERTIFICATE

NOTES FOR FORMS 1 AND 2 (from BS 7671)

1   The Electrical Installation Certificate is to be used only for the initial certification of a new installation or for an addition or alteration to an existing installation where new circuits have been introduced.

It is not to be used for a Periodic Inspection, for which an Electrical Installation Condition Report form should be used. For an addition or alteration which does not extend to the introduction of new circuits, a Minor Electrical Installation Works Certificate may be used.

The 'original' Certificate is to be given to the person ordering the work (Regulation 632.1). A duplicate should be retained by the contractor.

2   This Certificate is only valid if accompanied by the Schedule of Inspections and the Schedule(s) of Test Results.

3   The signatures appended are those of the persons authorized by the companies executing the work of design, construction, inspection and testing respectively. A signatory authorized to certify more than one category of work should sign in each of the appropriate places.

4   The time interval recommended before the first periodic inspection must be inserted (see IET Guidance Note 3 for guidance).

5   The page numbers for each of the Schedules of Test Results should be indicated, together with the total number of sheets involved.

6   The maximum prospective value of fault current (I_{pf}) recorded should be the greater of either the prospective value of short-circuit current or the prospective value of earth fault current.

7   The proposed date for the next inspection should take into consideration the frequency and quality of maintenance that the installation can reasonably be expected to receive during its intended life, and the period should be agreed between the designer, installer and other relevant parties.

GUIDANCE FOR RECIPIENTS (to be appended to the Certificate)

This safety Certificate has been issued to confirm that the electrical installation work to which it relates has been designed, constructed, inspected and tested in accordance with British Standard 7671 (the IET Wiring Regulations).

You should have received an 'original' Certificate and the contractor should have retained a duplicate. If you were the person ordering the work, but not the owner of the installation, you should pass this Certificate, or a full copy of it including the schedules, immediately to the owner.

The 'original' Certificate should be retained in a safe place and be shown to any person inspecting or undertaking further work on the electrical installation in the future. If you later vacate the property, this Certificate will demonstrate to the new owner that the electrical installation complied with the requirements of British Standard 7671 at the time the Certificate was issued. The Construction (Design and Management) Regulations require that, for a project covered by those Regulations, a copy of this Certificate, together with schedules, is included in the project health and safety documentation.

For safety reasons, the electrical installation will need to be inspected at appropriate intervals by a competent person. The maximum time interval recommended before the next inspection is stated on Page 1 under 'NEXT INSPECTION'.

This Certificate is intended to be issued only for a new electrical installation or for new work associated with an addition or alteration to an existing installation. It should not have been issued for the inspection of an existing electrical installation. An 'Electrical Installation Condition Report' should be issued for such an inspection.

Form 3   Form No: ..555513../3

SCHEDULE OF INSPECTIONS (for new installation work only)

Methods of protection against electric shock	Prevention of mutual detrimental influence

Both basic and fault protection:

N/A	(i)	SELV (note 1)
N/A	(ii)	PELV
N/A	(iii)	Double insulation
N/A	(iv)	Reinforced insulation

Basic protection: (note 2)

✓	(i)	Insulation of live parts
✓	(ii)	Barriers or enclosures
N/A	(iii)	Obstacles (note 3)
N/A	(iv)	Placing out of reach (note 4)

Fault protection:

(i) Automatic disconnection of supply:

✓	Presence of earthing conductor
✓	Presence of circuit protective conductors
✓	Presence of protective bonding conductors
N/A	Presence of supplementary bonding conductors
N/A	Presence of earthing arrangements for combined protective and functional purposes
N/A	Presence of adequate arrangements for other sources, where applicable
N/A	FELV
✓	Choice and setting of protective and monitoring devices (for fault and/or overcurrent protection)

(ii) Non-conducting location: (note 5)

N/A	Absence of protective conductors

(iii) Earth-free local equipotential bonding: (note 6)

N/A	Presence of earth-free local equipotential bonding

(iv) Electrical separation: (note 7)

N/A	Provided for **one item** of current-using equipment
N/A	Provided for **more than one item** of current-using equipment

Additional protection:

✓	Presence of residual current devices(s)
N/A	Presence of supplementary bonding conductors

Prevention of mutual detrimental influence

✓	(a)	Proximity to non-electrical services and other influences
✓	(b)	Segregation of Band I and Band II circuits or use of Band II insulation
✓	(c)	Segregation of safety circuits

Identification

✓	(a)	Presence of diagrams, instructions, circuit charts and similar information
✓	(b)	Presence of danger notices and other warning notices
✓	(c)	Labelling of protective devices, switches and terminals
✓	(d)	Identification of conductors

Cables and conductors

✓	Selection of conductors for current-carrying capacity and voltage drop
✓	Erection methods
✓	Routing of cables in prescribed zones
✓	Cables incorporating earthed armour or sheath, or run within an earthed wiring system, or otherwise adequately protected against nails, screws and the like
✓	Additional protection provided by 30 mA RCD for cables concealed in walls (where required in premises not under the supervision of a skilled or instructed person)
✓	Connection of conductors
✓	Presence of fire barriers, suitable seals and protection against thermal effects

General

✓	Presence and correct location of appropriate devices for isolation and switching
✓	Adequacy of access to switchgear and other equipment
N/A	Particular protective measures for special installations and locations
✓	Connection of single-pole devices for protection or switching in line conductors only
✓	Correct connection of accessories and equipment
N/A	Presence of undervoltage protective devices
✓	Selection of equipment and protective measures appropriate to external influences
✓	Selection of appropriate functional switching devices

Inspected by .... *G Wilson* ....   Date ..20-Jan-2012..

NOTES:

✓   to indicate an inspection has been carried out and the result is satisfactory
N/A   to indicate that the inspection is not applicable to a particular item
An entry must be made in every box.

1.   SELV An extra-low voltage system which is electrically separated from Earth and from other systems. The particular requirements of the Regulations must be checked (see Regulation 417.3)
2.   Method of basic protection - will include measurement of distances where appropriate
3.   Obstacles - only adopted in special circumstances (see Regulations 416.2 and 417.2)
4.   Placing out of reach - only adopted in special circumstances (see Regulation 417.3)

5.   Non-conducting locations - not applicable in domestic premises and requiring special precautions (see Regulation 418.1)
6.   Earth-free local equipotential bonding - not applicable in domestic premises, only used in special circumstances (see Regulation 418.2)
7.   Electrical separation (see Section 413 and Regulation 418.3)

Page .3 of .4

CITY OF LIVERPOOL COLLEGE
VAUXHALL ROAD
LIVERPOOL
L3 6BN

Form 4
Generic Schedule of Test Results for a single-phase installation

Form 4

GENERIC SCHEDULE OF TEST RESULTS

DB reference no	Consumer Unit
Location	Under-stairs cupboard
Zs at DB (Ω)	0.29
I_{pf} at DB (kA)	0.8 kA
Correct supply polarity confirmed	☑
Phase sequence confirmed (where appropriate)	N/A

Details of circuits and/or installed equipment vulnerable to damage when testing
E.L.V. lights in bathroom

Form No: 1235...../4

Details of test instruments used (state serial and/or asset numbers)
Continuity .... 1012F multi function
Insulation resistance .... "
Earth fault loop impedance .... "
RCD .... "
Earth electrode resistance .... N/A

Tested by:
Name (Capitals): G THOMPSON
Signature: G. Thompson   Date: 17-Jan-2012

Circuit number	Circuit Description	Overcurrent device BS (EN)	type	rating (A)	breaking capacity (kA)	Reference Method	Live (mm²)	cpc (mm²)	Ring final circuit continuity (Ω) r_1 (line)	r_n (neutral)	r_2 (cpc)	Continuity (Ω) (R_1+R_2) or R_2 *	R_2	Insulation Resistance (MΩ) Live-Live	Live-E	Polarity	Z_s (Ω)	RCD (ms) @ $I_{\Delta n}$	@ $5I_{\Delta n}$	Test button / functionality	Remarks (continue on a separate sheet if necessary)
	2	3	4	5	6	7	8	9	10	11	12	13	14	15	16	17	18	19	20	21	22
1	Ring- sockets downstairs	60898	B	32	6	100	2.5	1.5	0.62	0.62	1.02	0.41	N/A	+299	+299	✓	0.71	28	16	✓	NOTE: The high earth loop impedance
2	Ring - sockets upstairs	60898	B	32	6	100	2.5	1.5	0.62	0.62	1.02	0.41	N/A	+299	+299	✓	0.71	36	21	✓	on the upstairs lighting circuit was
3	Ring - kitchen	60898	B	32	6	100	2.5	1.5	0.22	0.22	0.37	0.15	N/A	+299	+299	✓	0.44	25	18	✓	found to be due to loose terminals
4	Cooker - kitchen	60898	B	32	6	100	6.0	2.5	N/A	N/A	N/A	0.16	N/A	+299	+299	✓	0.46	34	21	✓	at the point of connection of the
5	Lights - downstairs	60898	B	6	6	100	1.5	1.0	N/A	N/A	N/A	2.56	N/A	+299	+299	✓	2.85	29	19	✓	additional lighting circuitry.
6	Lights - upstairs	60898	B	6	6	100	1.5	1.0	N/A	N/A	N/A	8.20	N/A	+299	+299	✓	8.50	33	17	✓	With this additional circuitry
7	Lights - Garage	60898	B	6	6	100	1.5	1.0	N/A	N/A	N/A	1.51	N/A	+299	+299	✓	1.90	31	17	✓	disconnected, a satisfactory
8	Spare																				measurement of 2.86 ohms was
																					obtained. However, the circuit
																					conductors remain damaged and
																					affected cabling must be replaced.

Test results

* Where there are no spurs connected to a ring final circuit this value is also the $(R_1 + R_2)$ of the circuit.

Page 4 of 4

Form 4
Generic Schedule of Test Results for part of a three-phase installation

Form 4

Form No: ..555513...4

GENERIC SCHEDULE OF TEST RESULTS

DB reference no	DB1
Location	Main Switchroom
Zs at DB (Ω)	0.30
Ipf at DB (kA)	0.8
Correct supply polarity confirmed	☑
Phase sequence confirmed (where appropriate)	☑

Details of circuits and/or installed equipment vulnerable to damage when testing .... None

Details of test instruments used (state serial and/or asset numbers)
Continuity	1105M6E
Insulation resistance	As above
Earth fault loop impedance	As above
RCD	As above
Earth electrode resistance	N/A

Tested by:
Name (Capitals) ... G WILSON
Signature ... G Wilson   Date ... 20 Jan 2012

Circuit number	Circuit Description	Overcurrent device BS (EN)	type	rating (A)	breaking capacity (kA)	Reference Method	Live (mm²)	cpc (mm²)	Ring final circuit continuity (Ω) r₁ (line)	rₙ (neutral)	r₂ (cpc)	Continuity (Ω) (R₁ + R₂) * or R₂ (R₁ + R₂) *	R₂	Insulation Resistance (MΩ) Live-Live	Live-E	Polarity	Zₛ (Ω)	RCD (ms) @ IΔn	@ 5IΔn	Test button / functionality	Remarks (continue on a separate sheet if necessary)
	2	3	4	5	6	7	8	9	10	11	12	13	14	15	16	17	18	19	20	21	22
1	L1 Ring - Ground floor	60898	B	32	10	B	4.0	1.5	0.65	0.65	1.81	0.61	N/A	+299	+299	✓	0.91	52	15.3	✓	RCD 1
2	L1 Radial - Kitchen	60898	B	32	10	B	4.0	1.5	N/A	N/A	N/A	0.51	N/A	+299	+299	✓	0.81	"	"	✓	RCD 1
3	L1 Radial - Fire alarm	60898	B	10	10	B	2.5	1.5	N/A	N/A	N/A	0.58	N/A	+299	+299	✓	0.88				
4	L1 Lighting - ground fl	60898	B	16	10	A	2.5	1.5	N/A	N/A	N/A	1.08	N/A	+299	+299	✓	1.38				Cable tray + architrave switch drops
5	L2 Ring - Ground floor	60898	B	32	10	B	4.0	1.5	0.65	0.65	1.81	0.61	N/A	+299	+299	✓	0.91	51	15.1	✓	RCD 2
6	L2 Lighting - ground fl	60898	B	10	10	A	2.5	1.5	N/A	N/A	N/A	1.08	N/A	+299	+299	✓	1.38				Cable tray + architrave switch drops
7	L2 External lighting	60898	B	10	10	C	2.5	1.5	N/A	N/A	N/A	1.35	N/A	+299	+299	✓	1.65	51	15.1	✓	RCD 2
8	Spare																				
9	L3 Ring - ground floor	60898	B	32	10	B	4.0	1.5	0.65	0.65	1.82	0.61	N/A	+299	+299	✓	0.91	50	15.0	✓	RCD 3
10	L2 Radial - a/c plant room	60898	C	20	10	B	4.0	1.5	N/A	N/A	N/A	1.15	N/A	+299	+299	✓	1.36	"	"	✓	RCD 3
11	L2 Lighting - ground fl	60898	B	10	16	A	2.5	1.5	N/A	N/A	N/A	1.07	N/A	+299	+299	✓	1.37				Cable tray + architrave switch drops
12	Spare																				

Test results

* Where there are no spurs connected to a ring final circuit this value is also the (R₁ + R₂) of the circuit.

Page .4. of .4.

GENERIC SCHEDULE OF TEST RESULTS

NOTES

The following notes relate to the column number in the form (Form 4).

1 Circuit number, for three-phase installations it is preferred to use the designation L1, L2, L3 so (for example) for the 5th circuit, the designation 5L1, 5L2 and 5L3.

2 Circuit description, can be brief e.g. fluorescent lighting; see completed sheets.

3 BS (EN), enter the Standard of manufacture of the circuit protective device, e.g. (BS EN) 60898.

4 Type - where relevant for circuit-breakers enter the sensitivity type e.g. C.

5 Rating —enter the protective device current rating.

6 Breaking capacity – enter the protective device breaking capacity, often 'printed' on circuit-breakers, e.g. 6.

7 Reference Method – enter the cable's installed reference method, by using Table 4A2 of BS 7671

8 Conductor size – enter live conductor size in mm².

9 Conductor size – enter circuit protective conductor size in mm².

10 Ring line-line open resistance continuity in ohms, see 2.7.6.

11 Ring neutral-neutral open resistance continuity in ohms, see 2.7.6.

12 Ring cpc-cpc open resistance continuity in ohms, see 2.7.6.

13 Ring ($R_1 + R_2$) – enter the value recorded whilst carrying out Step 3 of the ring continuity test, see 2.7.6. Note that where senseless results are recorded, due to parallel return paths and it has been established and the inspector has verified continuity, a value is not necessary in this cell, and the cell may be ticked.

14 Continuity R2 – add the value of the cpc continuity reading. If using Test method 2, the 'wandering lead' method, then enter the maximum value of the various readings that were measured on the circuit. Note that where senseless results are recorded, due to parallel return paths and it has been established and the inspector has verified continuity, a value is not necessary in this cell, and the cell may be ticked.

15 Insulation, live—live – enter the minimum value recorded during testing the circuit for each of the various configurations, see 2.7.7.

16 Insulation, live—Earth – enter the minimum value recorded during testing the circuit for each of the various configurations, see 2.7.7.

17 Polarity – tick this cell when the polarity for the circuit has been confirmed, see 2.7.12.

18 Z_s – enter the circuit earth fault loop impedance by whatever method you have selected to determine it by, see 2.7.15.

19, 20 and 21 Enter the results from the tests carried out on any RCDs fitted to the circuit, see 2.7.18.

22 Remarks – this cell is provided to note anything relevant to the circuit and testing, see the completed examples of Form 4.

Form 5   Form No: ....1234.../5

MINOR ELECTRICAL INSTALLATION WORKS CERTIFICATE
(REQUIREMENTS FOR ELECTRICAL INSTALLATIONS - BS 7671 [IET WIRING REGULATIONS])
To be used only for minor electrical work which does not include the provision of a new circuit

---

PART 1:Description of minor works

1. Description of the minor works **2 new lighting points to home-office/bedroom 3 of dwelling.**

2. Location/Address **41 Larkspur Drive, Newtown. E. Sussex**

   Post Code **EA1 2BB**

3. Date minor works completed **27-Jan-2012**

4. Details of departures, if any, from BS 7671:2008, amended to ...**2011**..... (date)
   None. Electricity supplier's terminal equipment in need of attention. Cut-out fuse carrier cracked. Customer advised to contact supplier.

PART 2:Installation details

1. System earthing arrangement   TN-C-S ☑   TN-S ☐   TT ☐

2. Method of fault protection **ADS**

3. Protective device for the modified circuit   Type **BS EN 61009 Type B** Rating .....**6**... A

Comments on existing installation, including adequacy of earthing and bonding arrangements (see Regulation 132.16):
Existing circuit not provided with additional protection by RCD. Lighting circuit MCB converted to RCBO in order adequately to provide protection against damage to cables in walls; Reg. 522.6.101 refers.

PART 3:Essential Tests
Earth continuity satisfactory ☑

Insulation resistance:
   Line/neutral ....**+299**.. MΩ

   Line/earth ....**+299**.. MΩ

   Neutral/earth.............................**+299**.. MΩ

Earth fault loop impedance ....**1.2**.. Ω

Polarity satisfactory ☑

RCD operation (if applicable). Rated residual operating current $I_{\Delta n}$ ......**30**..mA and operating time of .....**28**..ms (at $I_{\Delta n}$)

PART 4:Declaration

I/We CERTIFY that the said works do not impair the safety of the existing installation, that the said works have been designed, constructed, inspected and tested in accordance with BS 7671:2008 (IET Wiring Regulations), amended to **2011**........ (date) and that the said works, to the best of my/our knowledge and belief, at the time of my/our inspection, complied with BS 7671 except as detailed in Part 1 above.

Name: **G Thompson**	Signature: *G Thompson*
For and on behalf of: **T and G Electrical**	Position: **Proprietor**
Address: **25 Whiteleaf Close**	
Newtown	Date: **27-Jan-2012**
E Sussex Post code **EA4 5XX**	

Page 1 of 1

CITY OF LIVERPOOL COLLEGE
VAUXHALL ROAD
LIVERPOOL
L3 6BN

MINOR ELECTRICAL INSTALLATION WORKS CERTIFICATE
NOTES ON COMPLETION

Scope

The Minor Works Certificate is intended to be used for alterations or additions to an installation that do not extend to the provision of a new circuit. Examples include the addition of socket-outlets or lighting points to an existing circuit, the relocation of a light switch etc. This Certificate may also be used for the replacement of equipment such as accessories or luminaires, but not for the replacement of distribution boards or similar items. Appropriate inspection and testing, however, should always be carried out irrespective of the extent of the work undertaken.

Part 1   Description of minor works

1,2 The minor works must be so described that the work that is the subject of the certification can be readily identified.

4   See Regulations 120.3 and 133.5. No departures are to be expected except in most unusual circumstances. See also Regulation 633.1.

Part 2   Installation details

2   The method of fault protection must be clearly identified, e.g. Automatic disconnection of supply (ADS).

If the existing installation lacks either an effective means of earthing or adequate main equipotential bonding conductors, this must be clearly stated in the space provided. See Regulation 633.2.

Recorded departures from BS 7671 may constitute non-compliance with the Electricity Safety, Quality and Continuity Regulations 2002 (as amended) or the Electricity at Work Regulations 1989. It is important that the client is advised immediately in writing.

Part 3   Essential tests

The relevant provisions of Part 6 'Inspection and testing' of BS 7671 must be applied in full to all minor works. For example, where a socket-outlet is added to an existing circuit it is necessary to:

1   establish that the earthing contact of the socket-outlet is connected to the main earthing terminal

2   measure the insulation resistance of the circuit that has been added to, and establish that it complies with Table 61 of BS 7671

3   measure the earth fault loop impedance to establish that the maximum permitted disconnection time is not exceeded

4   check that the polarity of the socket-outlet is correct

5   (if the work is protected by an RCD) verify the effectiveness of the RCD.

Part 4   Declaration

When making out and signing a form on behalf of a company or other business entity, individuals must state for whom they are acting.

GUIDANCE FOR RECIPIENTS (to be appended to the Certificate)

This Certificate has been issued to confirm that the electrical installation work to which it relates has been designed, constructed, inspected and tested in accordance with British Standard 7671 (*IET Wiring Regulations*).

You should have received an 'original' Certificate and the contractor should have retained a duplicate. If you were the person ordering the work, but not the owner of the installation, you should pass this Certificate, or a copy of it, to the owner.

A separate Certificate should have been received for each existing circuit on which minor works have been carried out. This Certificate is not appropriate if you requested the contractor to undertake more extensive installation work, for which you should have received an Electrical Installation Certificate.

The Certificate should be retained in a safe place and be shown to any person inspecting or undertaking further work on the electrical installation in the future. If you later vacate the property, this Certificate will demonstrate to the new owner that the minor electrical installation work carried out complied with the requirements of British Standard 7671 at the time the Certificate was issued.

Form 6

Form No: **1235**......./6

ELECTRICAL INSTALLATION CONDITION REPORT

SECTION A. DETAILS OF THE CLIENT / PERSON ORDERING THE REPORT

Name M Parker

Address The Beeches

Quaintlife

E Sussex

Post Code: EA11 2ZZ

SECTION B. REASON FOR PRODUCING THIS REPORT Client reported flickering lights

Date(s) on which inspection and testing was carried out

SECTION C. DETAILS OF THE INSTALLATION WHICH IS THE SUBJECT OF THIS REPORT

Occupier As above

Address As above

Post Code: EA11 2ZZ

Description of premises (tick as appropriate)

Domestic ☑ Commercial ☐ Industrial ☐ Other (include brief description) ☐

Estimated age of wiring system ....9...years

Evidence of additions / alterations Yes ☑ No ☐ Not apparent ☐ If yes, estimate age ....4....years

Installation records available? (Regulation 621.1) Yes ☐ No ☑   Date of last inspection .... (date)

SECTION D. EXTENT AND LIMITATIONS OF INSPECTION AND TESTING

Extent of the electrical installation covered by this report

Visual inspection only of suppliers terminal equipment, inspection and test of consumer unit, main protective bonding conductors, supplementary bonding conductors and final circuits

Agreed limitations including the reasons (see Regulation 634.2) No disturbance of building fabric

Agreed with: Client

Operational limitations including the reasons (see page no............) None

The inspection and testing detailed in this report and accompanying schedules have been carried out in accordance with BS 7671: 2008 (IET Wiring Regulations) as amended to .2011.........................

It should be noted that cables concealed within trunking and conduits, under floors, in roof spaces, and generally within the fabric of the building or underground, have **not** been inspected unless specifically agreed between the client and inspector prior to the inspection.

SECTION E. SUMMARY OF THE CONDITION OF THE INSTALLATION

General condition of the installation (in terms of electrical safety) This installation was constructed when the requirements of BS7671: 2001) were in place. The installation does not comply with the requirements of BS7671: 2008 (Amendment No1) but is in generally good condition, apart from the connection arrangement of additional lighting points which were installed in 2008. This has given rise to the flickering lights and presents a potential risk of fire and electric shock.

Overall assessment of the installation in terms of its suitability for continued use

~~SATISFACTORY~~ / UNSATISFACTORY* (Delete as appropriate)

*An unsatisfactory assessment indicates that dangerous (code C1) and/or potentially dangerous (code C2) conditions have been identified.

SECTION F. RECOMMENDATIONS

Where the overall assessment of the suitability of the installation for continued use above is stated as UNSATISFACTORY, I / we recommend that any observations classified as *'Danger present'* (code C1) or *'Potentially dangerous'* (code C2) are acted upon as a matter of urgency. Investigation without delay is recommended for observations identified as *'further investigation required'.*

Observations classified as *'Improvement recommended'* (code C3) should be given due consideration.

Subject to the necessary remedial action being taken, I / we recommend that the installation is further inspected and tested by Jan 2015 (date)

SECTION G. DECLARATION

I/We, being the person(s) responsible for the inspection and testing of the electrical installation (as indicated by my/our signatures below), particulars of which are described above, having exercised reasonable skill and care when carrying out the inspection and testing, hereby declare that the information in this report, including the observations and the attached schedules, provides an accurate assessment of the condition of the electrical installation taking into account the stated extent and limitations in section D of this report.

Inspected and tested by:	Report authorised for issue by:
Name (Capitals) G THOMPSON	Name (Capitals) G THOMPSON
Signature *G Thompson*	Signature *G Thompson*
For/on behalf of T and C ELECTRICAL	For/on behalf of T and C Electrical
Position Proprietor	Position Proprietor
Address 25 Whiteleaf Close, Newtown, E Sussex	Address 25 Whiteleaf Close, Newtown, E Sussex
Post code EA4 5XX Date 17-Jan-2012	Post code EA4 5XX Date 17-Jan-2012

SECTION H. SCHEDULE(S)

.......1......schedule(s) of inspection and ....1.....schedule(s) of test results are attached.

The attached schedule(s) are part of this document and this report is valid only when they are attached to it.

Page 1 of 2

Form 6   Form No: 1235......./6

SECTION I. SUPPLY CHARACTERISTICS AND EARTHING ARRANGEMENTS			
Earthing arrangements	**Number and Type of Live Conductors**	**Nature of Supply Parameters**	**Supply Protective Device**
TN-C ☐ TN-S ☐ TN-C-S ☑ TT ☐ IT ☐	a.c. ☑ d.c. ☐ 1-phase, 2-wire ☑ 2-wire ☐ 2 phase, 3-wire ☐ 3-wire ☐ 3 phase, 3-wire ☐ Other ☐ 3 phase, 4-wire ☐ Confirmation of supply polarity ☑	Nominal voltage, U / U$_0$(1) ....230..V Nominal frequency, f(1) ....50....Hz Prospective fault current, I$_{pf}$(2) ....0.8..kA External loop impedance, Ze(2) .....0.29..Ω Note: (1) by enquiry (2) by enquiry or by measurement	BS (EN) BS 1361 Fuse Type 2 Rated current ......100...A

Other sources of supply (as detailed on attached schedule) ☐

SECTION J. PARTICULARS OF INSTALLATION REFERRED TO IN THE REPORT

Means of Earthing	Details of Installation Earth Electrode (where applicable)
Distributor's facility ☑ Installation earth electrode ☐	Type ......N/A......... Location N/A......... Resistance to Earth ......N/A.Ω

Main Protective Conductors

Earthing conductor	Material Copper	csa ....16.....mm²	Connection / continuity verified ☑
Main protective bonding conductors	Material Copper	csa ....16.....mm²	Connection / continuity verified ☑

To incoming water service ☑   To incoming gas service ☑   To incoming oil service ☐   To structural steel ☐

To lightning protection ☐   To other incoming service(s) ☐ Specify ....

Main Switch / Switch-Fuse / Circuit-Breaker / RCD

Location Under-stairs cupboard .... BS(EN) 60947-3 No of poles 2	Current rating ....100.A Fuse / device rating or setting N/A.A Voltage rating 230.V	If RCD main switch Rated residual operating current (I$_{\Delta n}$) ....N/A.mA Rated time delay ....N/A.ms Measured operating time(at I$_{\Delta n}$) ....N/A.ms

SECTION K. OBSERVATIONS

Referring to the attached schedules of inspection and test results, and subject to the limitations specified at the *Extent and limitations of inspection and testing* section

No remedial action is required ☐   The following observations are made ☑ (see below):

OBSERVATION(S)	CLASSIFICATION CODE	FURTHER INVESTIGATION REQUIRED (YES / NO)
1. Damage to lighting circuit at junction box. Arcing and burnt insulation evident.	C1	No
2. Loose connection as above, causing unreliable earthing of circuit and risk of fire.	C2	No
3. No additional protection by RCD for low voltage circuits in bathroom	C3	No
4. No earthed, mechanical or additional protection by RCD for cables concealed in walls.	C3	No
NOTE: Faulty section of lighting circuit disconnected with client's approval.		

One of the following codes, as appropriate, has been allocated to each of the observations made above to indicate to the person(s) responsible for the installation the degree of urgency for remedial action.

C1 – Danger present. Risk of injury. Immediate remedial action required

C2 – Potentially dangerous - urgent remedial action required

C3 – Improvement recommended

Page 2 of .2.

Form 7   Form No: 1235......./7

CONDITION REPORT INSPECTION SCHEDULE FOR
DOMESTIC AND SIMILAR PREMISES WITH UP TO 100 A SUPPLY

Note: This form is suitable for many types of smaller installation not exclusively domestic.

OUTCOMES	Acceptable condition	✓	Unacceptable condition	State **C1** or **C2**	Improvement recommended	State **C3**	Not verified **N/V**	Limitation **LIM**	Not applicable **N/A**

ITEM NO	DESCRIPTION	OUTCOME (Use codes above. Provide additional comment where appropriate. C1, C2 and C3 coded items to be recorded in Section K of the Condition Report)	Further investigation required? (*Y* or *N*)
1.0	**DISTRIBUTOR'S / SUPPLY INTAKE EQUIPMENT**		
1.1	Service cable condition	N/A	No
1.2	Condition of service head	✓	No
1.3	Condition of tails - Distributor	✓	No
1.4	Condition of tails - Consumer	✓	No
1.5	Condition of metering equipment	✓	No
1.6	Condition of isolator (where present)	N/A	No
2.0	**PRESENCE OF ADEQUATE ARRANGEMENTS FOR OTHER SOURCES SUCH AS MICROGENERATORS (551.6; 551.7)**	N/A	No
3.0	**EARTHING / BONDING ARRANGEMENTS (411.3; Chap 54)**		
3.1	Presence and condition of distributor's earthing arrangement (542.1.2.1; 542.1.2.2)	✓	No
3.2	Presence and condition of earth electrode connection where applicable (542.1.2.3)	N/A	No
3.3	Provision of earthing / bonding labels at all appropriate locations (514.13)	✓	No
3.4	Confirmation of earthing conductor size (542.3; 543.1.1)	✓	No
3.5	Accessibility and condition of earthing conductor at MET (543.3.2)	✓	No
3.6	Confirmation of main protective bonding conductor sizes (544.1)	✓	No
3.7	Condition and accessibility of main protective bonding conductor connections (543.3.2; 544.1.2)	✓	No
3.8	Accessibility and condition of all protective bonding connections (543.3.2)	✓	No
4.0	**CONSUMER UNIT(S) / DISTRIBUTION BOARD(S)**		
4.1	Adequacy of working space / accessibility to consumer unit / distribution board (132.12; 513.1)	✓	No
4.2	Security of fixing (134.1.1)	✓	No
4.3	Condition of enclosure(s) in terms of IP rating etc (416.2)	✓	No
4.4	Condition of enclosure(s) in terms of fire rating etc (526.5)	✓	No
4.5	Enclosure not damaged/deteriorated so as to impair safety (621.2(iii))	✓	No
4.6	Presence of main linked switch (as required by 537.1.4)	✓	No
4.7	Operation of main switch (functional check) (612.13.2)	✓	No
4.8	Manual operation of circuit-breakers and RCDs to prove disconnection (612.13.2)	✓	No
4.9	Correct identification of circuit details and protective devices (514.8.1; 514.9.1)	✓	No
4.10	Presence of RCD quarterly test notice at or near consumer unit / distribution board (514.12.2)	✓	No
4.11	Presence of non-standard (mixed) cable colour warning notice at or near consumer unit / distribution board (514.14)	✓	No
4.12	Presence of alternative supply warning notice at or near consumer unit / distribution board (514.15)	✓	No
4.13	Presence of other required labelling (please specify) (Section 514)	✓	No
4.14	Examination of protective device(s) and base(s); correct type and rating (no signs of unacceptable thermal damage, arcing or overheating) (421.1.3)	✓	No
4.15	Single-pole protective devices in line conductor only (132.14.1; 530.3.2)	✓	No
4.16	Protection against mechanical damage where cables enter consumer unit / distribution board (522.8.1; 522.8.11)	N/A	No
4.17	Protection against electromagnetic effects where cables enter consumer unit / distribution board / enclosures (521.5.1)	N/A	No
4.18	RCD(s) provided for fault protection – includes RCBOs (411.4.9; 411.5.2; 531.2)	N/A	No
4.19	RCD(s) provided for additional protection - includes RCBOs (411.3.3; 415.1)	C3	No

Page .1. of .2.

| | Form 7 | | | | | Form No: 1235......./7 | | | | |

OUTCOMES	Acceptable condition	✓	Unacceptable condition	State **C1** or **C2**	Improvement recommended	State **C3**	Not verified **N/V**	Limitation **LIM**	Not applicable **N/A**

ITEM NO	DESCRIPTION	OUTCOME (Use codes above. Provide additional comment where appropriate. C1, C2 and C3 coded items to be recorded in Section K of the Condition Report)	Further investigation required? (**Y** or **N**)
5.0	**FINAL CIRCUITS**		
5.1	Identification of conductors (514.3.1)	✓	No
5.2	Cables correctly supported throughout their run (522.8.5)	✓	No
5.3	Condition of insulation of live parts (416.1)	C1	No
5.4	Non-sheathed cables protected by enclosure in conduit, ducting or trunking (521.10.1)	✓	No
	▪ To include the integrity of conduit and trunking systems (metallic and plastic)	✓	No
5.5	Adequacy of cables for current-carrying capacity with regard for the type and nature of installation (Section 523)	✓	No
5.6	Coordination between conductors and overload protective devices (433.1; 533.2.1)	✓	No
5.7	Adequacy of protective devices: type and rated current for fault protection (411.3)	✓	No
5.8	Presence and adequacy of circuit protective conductors (411.3.1.1; 543.1)	✓	No
5.9	Wiring system(s) appropriate for the type and nature of the installation and external influences (Section 522)	✓	No
5.10	Concealed cables installed in prescribed zones (see Section D. *Extent and limitations*) (522.6.101)	✓	No
5.11	Concealed cables incorporating earthed armour or sheath, or run within earthed wiring system, or otherwise protected against mechanical damage from nails, screws and the like (see Section D. *Extent and limitations*) (522.6.101; 522.6.103)	✓	No
5.12	Provision of additional protection by RCD not exceeding 30 mA:		
	▪ for all socket-outlets of rating 20 A or less provided for use by ordinary persons unless an exception is permitted (411.3.3)	✓	No
	▪ for supply to mobile equipment not exceeding 32 A rating for use outdoors (411.3.3)		No
	▪ for cables concealed in walls or partitions (522.6.102; 522.6.103)	C3	No
5.13	Provision of fire barriers, sealing arrangements and protection against thermal effects (Section 527)	✓	No
5.14	Band II cables segregated / separated from Band I cables (528.1)	N/A	No
5.15	Cables segregated / separated from communications cabling (528.2)	N/A	No
5.16	Cables segregated / separated from non-electrical services (528.3)	✓	No
5.17	Termination of cables at enclosures – indicate extent of sampling in Section D of the report (Section 526)		
	▪ Connections soundly made and under no undue strain (526.6)	C1	No
	▪ No basic insulation of a conductor visible outside enclosure (526.8)	✓	No
	▪ Connections of live conductors adequately enclosed (526.5)	✓	No
	▪ Adequately connected at point of entry to enclosure (glands, bushes etc.) (522.8.5)	✓	No
5.18	Condition of accessories including socket-outlets, switches and joint boxes (621.2(iii))	C1	No
5.19	Suitability of accessories for external influences (512.2)	✓	No

6.0	**LOCATION(S) CONTAINING A BATH OR SHOWER**		
6.1	Additional protection for all low voltage (LV) circuits by RCD not exceeding 30 mA (701.411.3.3)	C3	No
6.2	Where used as a protective measure, requirements for SELV or PELV met (701.414.4.5)	N/A	No
6.3	Shaver sockets comply with BS EN 61558-2-5 formerly BS 3535 (701.512.3)	✓	No
6.4	Presence of supplementary bonding conductors, unless not required by BS 7671:2008 (701.415.2)	✓	No
6.5	Low voltage (e.g. 230 volt) socket-outlets sited at least 3 m from zone 1 (701.512.3)	✓	No
6.6	Suitability of equipment for external influences for installed location in terms of IP rating (701.512.2)	✓	No
6.7	Suitability of equipment for installation in a particular zone (701.512.3)	✓	No
6.8	Suitability of current-using equipment for particular position within the location (701.55)	✓	No

7.0	**OTHER PART 7 SPECIAL INSTALLATIONS OR LOCATIONS**		
7.1	List all other special installations or locations present, if any. (Record separately the results of particular inspections applied.)	N/A	No

Inspected by:
Name (Capitals) **G THOMPSON** .... Signature ...... *G. Thompson* .... Date 17-Jan-2012

Page .2. of .2.

ELECTRICAL INSTALLATION CONDITION REPORT

NOTES ON COMPLETION

1 This Report should only be used for reporting on the condition of an existing electrical installation. An installation which was designed to an earlier edition of the Regulations and which does not fully comply with the current edition is not necessarily unsafe for continued use, or requires upgrading. Only damage, deterioration, defects, dangerous conditions and non-compliance with the requirements of the Regulations, which may give rise to danger, should be recorded.

2 The Report, normally comprising at least six pages, should include schedules of both the inspection and the test results. Additional pages may be necessary for other than a simple installation. The number of each page should be indicated, together with the total number of pages involved.

3 The reason for producing this Report, such as change of occupancy or landlord's periodic maintenance, should be identified in Section B.

4 Those elements of the installation that are covered by the Report and those that are not should be identified in Section D (Extent and limitations). These aspects should have been agreed with the person ordering the report and other interested parties before the inspection and testing commenced. Any operational limitations, such as inability to gain access to parts of the installation or an item of equipment, should also be recorded in Section D.

5 The maximum prospective value of fault current (I_{pf}) recorded should be the greater of either the prospective value of short-circuit current or the prospective value of earth fault current.

6 Where an installation has an alternative source of supply a further schedule of supply characteristics and earthing arrangements based upon Section I of this Report should be provided.

7 A summary of the condition of the installation in terms of safety should be clearly stated in Section E. Observations, if any, should be categorised in Section K using the coding C1 to C3 as appropriate. Any observation given a code C1 or C2 classification should result in the overall condition of the installation being reported as unsatisfactory.

8 Wherever practicable, **items classified as 'Danger present' (C1) should be made safe on discovery**. Where this is not practical the owner or user should be given written notification as a matter of urgency.

9 Where an observation requires further investigation because the inspection has revealed an apparent deficiency which could not, owing to the extent or limitations of the inspection, be fully identified, this should be indicated in the column headed 'Further investigation required' within Section K.

10 If the space available for observations in Section K is insufficient, additional numbered pages should be provided as necessary.

11 The date by which the next Electrical Installation Condition Report is recommended should be given in Section F. The interval between inspections should take into account the type and usage of the installation and its overall condition.

GUIDANCE FOR RECIPIENTS (to be appended to the Report)

This Report is an important and valuable document which should be retained for future reference.

1   The purpose of this Condition Report is to confirm, so far as reasonably practicable, whether or not the electrical installation is in a satisfactory condition for continued service (see Section E). The Report should identify any damage, deterioration, defects and/or conditions which may give rise to danger (see Section K).

2   The person ordering the Report should have received the 'original' Report and the inspector should have retained a duplicate.

3   The 'original' Report should be retained in a safe place and be made available to any person inspecting or undertaking work on the electrical installation in the future. If the property is vacated, this Report will provide the new owner/occupier with details of the condition of the electrical installation at the time the Report was issued.

4   Where the installation incorporates a residual current device (RCD) there should be a notice at or near the device stating that it should be tested quarterly. **For safety reasons it is important that this instruction is followed.**

5   Section D (Extent and Limitations) should identify fully the extent of the installation covered by this Report and any limitations on the inspection and testing. The inspector should have agreed these aspects with the person ordering the Report and with other interested parties (licensing authority, insurance company, mortgage provider and the like) before the inspection was carried out.

6   Some operational limitations such as inability to gain access to parts of the installation or an item of equipment may have been encountered during the inspection. The inspector should have noted these in Section D.

7   For items classified in Section K as C1 ('Danger present'), **the safety of those using the installation is at risk,** and it is recommended that a competent person undertakes the necessary remedial work immediately.

8   For items classified in Section K as C2 ('Potentially dangerous'), **the safety of those using the installation may be at risk** and it is recommended that a competent person undertakes the necessary remedial work as a matter of urgency.

9   Where it has been stated in Section K that an observation requires further investigation the inspection has revealed an apparent deficiency which could not, due to the extent or limitations of the inspection, be fully identified. Such observations should be investigated as soon as possible. A further examination of the installation will be necessary, to determine the nature and extent of the apparent deficiency (see Section F).

10 For safety reasons, the electrical installation should be re-inspected at appropriate intervals by a competent person. The recommended date by which the next inspection is due is stated in Section F of the Report under 'Recommendations' and on a label at or near to the consumer unit/distribution board.

CONDITION REPORT INSPECTION SCHEDULE

GUIDANCE FOR INSPECTORS

1 Section 1.0. Where inadequacies in the distributor's equipment are encountered the inspector should advise the person ordering the work to inform the appropriate authority.

2 Older installations designed prior to BS 7671:2008 may not have been provided with RCDs for additional protection. The absence of such protection should as a minimum be given a code C3 classification (item 5.12).

3 The schedule is not exhaustive.

4 Numbers in brackets are Regulation references to specified requirements.

Appendix A
Maximum permissible measured earth fault loop impedance

A1   Tables

612.9
411.4.6
411.4.7
411.4.8 The tables in this appendix provide maximum permissible measured earth fault loop impedances (Z_s) for compliance with BS 7671. The values are those that must not be exceeded when the tests are carried out at an ambient temperature of 10 °C. Table A6 provides correction factors for other ambient temperatures.

Where the cables to be used are to Table 4, 7 or 8 of BS 6004 or Table 3, 5, 6 or 7 of BS 7211 or are other thermoplastic (PVC) or thermosetting (low smoke halogen-free – LSHF) cables to these British Standards, and the cable loading is such that the maximum operating temperature is 70 °C, then Tables A1 to A3 give the maximum earth loop impedances for circuits with:

1   protective conductors of copper and having from 1 mm² to 16 mm² cross-sectional area

2   an overcurrent protective device (i.e. a fuse) to BS 3036, BS 88-2 or BS 88-3.

For each type of fuse, two tables are given:

411.3.2.2   ▶   where the circuit concerned supplies final circuits not exceeding 32 A and the maximum disconnection time for compliance with Regulation 411.3.2.2 is 0.4 s for TN systems, and

411.3.2.3   ▶   where the circuit concerned is a final circuit exceeding 32 A or a distribution circuit and the disconnection time for compliance with Regulation 411.3.2.3 is 5 s for TN systems.

In each table the earth fault loop impedances given correspond to the appropriate disconnection time from a comparison of the time/current characteristics of the device 543.1.3 concerned and the adiabatic equation given in Regulation 543.1.3.

The tabulated values apply only when the nominal voltage to Earth (U_0) is 230 V.

Table A4 gives the maximum measured Z_s for circuits protected by circuit-breakers to BS 3871-1 and BS EN 60898, and RCBOs to BS EN 61009.

Tables 41.2 to 41.4   **Note:** The impedances tabulated in this appendix are lower than those in Tables 41.2 to 41.4 of BS 7671 as the impedances in this appendix are measured values at an assumed conductor temperature of 10 °C whilst those in BS 7671 are design figures at the conductor normal operating temperature. The correction factor (divisor) used is 1.24. For smaller section cables the impedance may also be limited by the adiabatic equation of Regulation 543.1.3 543.1.3. A value of k of 115 from Table 54.3 of BS 7671 is used. This is suitable for PVC insulated and sheathed cables to Table 4, 7 or 8 of BS 6004 and for thermosetting (LSHF) insulated and sheathed cables to Table 3, 5, 6 or 7 of BS 7211. The k value is based on both the thermoplastic (PVC) and thermosetting (LSHF) cables operating at a maximum temperature of 70 °C. The IET *Commentary on the Wiring Regulations* provides a full explanation.

▼ **Table A1** Semi-enclosed fuses. Maximum measured earth fault loop impedance (in ohms) at ambient temperature of 10 °C where the overcurrent protective device is a semi-enclosed fuse to BS 3036

(i) 0.4 second disconnection (final circuits not exceeding 32 A in TN systems)

Protective conductor (mm²)	Fuse rating (A)			
	5	15	20	30
1.0	7.7	2.1	1.4	NP
≥ 1.5	7.7	2.1	1.4	0.9

(ii) 5 seconds disconnection (final circuits exceeding 32 A and distribution circuits in TN systems)

Protective conductor (mm²)	Fuse rating (A)			
	20	30	45	60
1.0	2.7	NP	NP	NP
1.5	3.1	2.0	NP	NP
2.5	3.1	2.1	1.2	NP
4.0	3.1	2.1	1.3	0.8
≥ 6.0	3.1	2.1	1.3	0.9

Note: NP means that the combination of the protective conductor and the fuse is Not Permitted.

▼ **Table A2** BS 88-2 fuses. Maximum measured earth fault loop impedance (in ohms) at ambient temperature of 10 °C where the overcurrent protective device is a fuse to BS 88-2

(i) 0.4 second disconnection (final circuits not exceeding 32 A in TN systems)

Protective conductor (mm²)	Fuse rating (A)					
	6	10	16	20	25	32
1.0	6.6	3.9	2.1	1.4	1.1	0.66
1.5	6.6	3.9	2.1	1.4	1.1	0.84
≥ 2.5	6.6	3.9	2.1	1.4	1.1	0.84

(ii) 5 seconds disconnection (final circuits exceeding 32 A and distribution circuits in TN systems)

Protective conductor (mm²)	Fuse rating (A)							
	20	25	32	40	50	63	80	100
1.0	1.7	1.2	0.66	NP	NP	NP	NP	NP
1.5	2.4	1.7	1.1	0.64	NP	NP	NP	NP
2.5	2.4	1.8	1.5	0.93	0.55	0.34	NP	NP
4.0	2.4	1.8	1.5	1.1	0.77	0.50	0.23	NP
6.0	2.4	1.8	1.5	1.1	0.84	0.66	0.36	0.22
10.0	2.4	1.8	1.5	1.1	0.84	0.66	0.46	0.32
16.0	2.4	1.8	1.5	1.1	0.84	0.66	0.46	0.37

Note: NP means that the combination of the protective conductor and the fuse is Not Permitted.

▼ **Table A3** BS 88-3 fuses. Maximum measured earth fault loop impedance (in ohms) at ambient temperature of 10 °C where the overcurrent protective device is a fuse to BS 88-3

(i) 0.4 second disconnection (final circuits not exceeding 32 A in TN systems)

Protective conductor (mm²)	Fuse rating (A)			
	5	16	20	30
1.0	8.4	1.9	1.6	0.64
1.5 to 16	8.4	1.9	1.6	0.77

(ii) 5 seconds disconnection (final circuits exceeding 32 A and distribution circuits in TN systems)

Protective conductor (mm²)	Fuse rating (A)					
	20	32	45	63	80	100
1.0	2.3	0.64	NP	NP	NP	NP
1.5	2.7	0.89	NP	NP	NP	NP
2.5	2.7	1.3	0.64	0.24	NP	NP
4.0	2.7	1.3	0.84	0.43	0.24	0.16
6.0	2.7	1.3	0.84	0.58	0.32	0.24
10	2.7	1.3	0.84	0.58	0.43	0.32
16	2.7	1.3	0.84	0.58	0.43	0.32

Note: NP means that the combination of the protective conductor and the fuse is Not Permitted.

▼ **Table A4** Circuit-breakers. Maximum measured earth fault loop impedance (in ohms) at ambient temperature of 10 °C where the overcurrent device is a circuit-breaker to BS 3871 or BS EN 60898 or RCBO to BS EN 61009

For 0.1 to 5 second disconnection times

Circuit-breaker type	Circuit-breaker rating (A)														
	5	6	10	15	16	20	25	30	32	40	45	50	63	100	125
1	9.27	7.73	4.64	3.09	2.90	2.32	1.85	1.55	1.45	1.16	1.03	0.93	0.74	0.46	0.37
2	5.3	4.42	2.65	1.77	1.66	1.32	1.06	0.88	0.83	0.66	0.59	0.53	0.42	0.26	0.21
B	7.42	6.18	3.71	2.47	2.32	1.85	1.48	1.24	1.16	0.93	0.82	0.74	0.59	0.37	0.30
3&C	3.71	3.09	1.85	1.24	1.16	0.93	0.74	0.62	0.58	0.46	0.41	0.37	0.29	0.19	0.15
D	1.85	1.55	0.93	0.62	0.58	0.46	0.37	0.31	0.29	0.23	0.21	0.19	0.15	0.09	0.07

Regulation 434.5.2 of BS 7671:2008 requires the protective conductor csa to meet the requirements of BS EN 60898-1, BS EN 60898-2 or BS EN 61009-1, or the minimum quoted by the manufacturer. Table A5 gives minimum protective conductor sizes for energy-limiting class 3 Types B and C devices only.

▼ **Table A5**   Minimum protective conductor size for class 3 Types B and C devices

Energy-limiting class 3 device rating (A)	Fault level (kA)	Protective conductor csa (mm²)*	
		Type B	Type C
≤16	≤3	1.0	1.5
≤16	≤6	2.5	2.5
16 < A ≤ 32	≤3	1.5	1.5
16 < A ≤ 32	≤6	2.5	2.5
40	≤3	1.5	1.5
40	≤6	2.5	2.5

* For other device types and ratings or higher fault levels consult manufacturer's data. See Regulation 434.5.2 and the IET publication *Commentary on the IEE Wiring Regulations*.

▼ **Table A6**   Ambient temperature correction factors

Ambient temperature (°C)	Correction factor (from 10 °C) (notes 1 and 2)
0	0.96
5	0.98
10	1.00
20	1.04
25	1.06
30	1.08

Notes:

1 The correction factor is given by: {1 + 0.004 (ambient temp − 10 °C} where 0.004 is the simplified resistance coefficient per °C at 20 °C given by BS EN 60228 for both copper and aluminium conductors.

2 The factors are different to those of Table B.2 because Table A6 corrects from 10 °C and Table B.2 from 20 °C.

The ambient correction factor of Table A6 is applied to the earth fault loop impedances of Tables A1 to A4 if the ambient temperature is other than 10 °C.

For example, if the ambient temperature is 25 °C the measured earth fault loop impedance of a circuit protected by a 32 A type B circuit-breaker to BS EN 60898 should not exceed $1.16 \times 1.06 = 1.23\ \Omega$.

A2   Appendix 14 of BS 7671:2008

Appx 14   Appendix 14 of BS 7671 takes into account the increase of the conductor resistance with increase of temperature due to load current, which may be used to verify compliance with the requirements of Regulation 411.4 for TN systems.

The requirements of Regulation 411.4.5 are considered met when the measured value of fault loop impedance satisfies the following equation:

$$Z_s(m) \le 0.8 \times \frac{U_0}{I_a}$$

where:

$Z_s(m)$ is the measured impedance of the fault current loop starting and ending at the point of fault (Ω)

U_0 is the nominal a.c. rms line voltage to earth (V)

I_a is the current causing the automatic operation of the protective device within the time stated in Table 41.1 or within 5 s according to the conditions stated in Regulation 411.3.2.3 (A).

A3   Methods of adjusting tabulated values of Z_s

(See also 2.7.15 'Earth fault loop impedance verification'.)

A circuit is wired in flat twin and cpc 70 °C thermoplastic (PVC) cable and protected by a 6 amp type B circuit-breaker to BS EN 60898. When tested at an ambient temperature of 20 °C, determine the maximum acceptable measured value of Z_s for the circuit.

Solution:

$$Z_{test\ (max)} = \frac{1}{F}Z_s$$

From Table 41.3(a) of BS 7671, the maximum permitted value of $Z_s = 7.67$ ohms

From Table B3 in Appendix B of this Guidance Note, $F = 1.20$

$$Z_{test\ (max)} = \frac{1}{1.20} \times 7.67$$

$$Z_{test\ (max)} = 6.39\ \text{ohms}$$

A more accurate value can be obtained if the external earth fault loop impedance, Z_e, is known. In this case, the following formula may be used:

$$Z_{test} \leq \frac{1}{F}\{Z_s + Z_e(F - 1)\}$$

In the example above, assume Z_e is 0.35. Thus, the more accurate value is:

$$Z_{test\ (max)} = \frac{1}{1.20} \times \{7.67 + 0.35(1.20 - 1)\}$$

$$Z_{test\ (max)} = 6.45\ \text{ohms}$$

Where the test ambient temperature is likely to be other than 20 °C, a further correction can be made to convert the value to the expected ambient temperature, using the following formula:

$$Z_{test\ (max)} = \frac{1}{F + 1 - \alpha}\{Z_s + Z_e(F - \alpha)\}$$

where α is given by Table B2 of Appendix B.

In the example above, assume the test ambient temperature is 5 °C.

From Table B2, $\alpha = 0.94$

Thus, the accurate reading including temperature compensation is:

$$Z_{test\ (max)} = \frac{1}{1.20 + 1 - 0.94}\{7.67 + 0.35(1.20 - 0.94)\}$$

$$Z_{test\ (max)} = 0.794\{7.67 + 0.091\}$$

$$Z_{test\ (max)} = 6.16\ ohms$$

Alternatively, a conductor temperature resistance factor, F, of 1.26, which corresponds to 5 °C, can be used instead of the 1.20 factor in the formula shown in the second solution box. This gives the same result as that shown above.

Note: If reduced cross-sectional area protective conductors are used, maximum earth fault loop impedances may need to be further reduced to ensure disconnection times are sufficiently short to prevent overheating of protective conductors during earth faults. The requirement of the equation in Regulation 543.1.3 needs to be met:

$$S \geq \frac{\sqrt{I^2\ t}}{k}$$

where:

S   is the nominal cross-sectional area of the conductor in mm^2

$\geq$ means greater than or equal to

k   is a factor from Tables 54.2–54.4

I   is the prospective earth fault current given by U_0/Z_s

Z_s is the loop impedance at conductor normal operating temperature

t   is the operating time of the overcurrent device in seconds. This is obtained from the graphs in Appendix 3 of BS 7671, as the prospective earth fault current $I (= U_0/Z_s)$ is known.

The following example illustrates how measurements taken at 20 °C may be adjusted to 70 °C values, taking the $(R_1 + R_2)$ reading for the circuit into account.

In the previous example, taking the $(R_1 + R_2)$ reading for the circuit as 0.2 ohm:

Z_s for the circuit at 70 °C

$$= Z_e + F(R_1 + R_2)_{test}$$

$$= 0.35 + 1.20 \times 0.2$$

$$= 0.59\ ohm$$

The temperature-corrected Z_s figure of 0.59 ohm is acceptable, since it is less than the maximum value of 7.67 ohms given in Table 41.3 of BS 7671.

The formula above involves taking measurements at 20 °C and converting them to 70 °C values. Alternatively, the 70 °C values can be converted to the values at the expected ambient temperature, e.g. 20 °C, when the measurement is carried out.

Taking the same circuit,

$$Z_{test} = Z_e + (R_1 + R_2)_{test}$$

$$= 0.35 + 0.2$$

$$= 0.55 \text{ ohm}$$

From the formula $Z_{test (max)} = \dfrac{1}{F} Z_s$

$Z_{s (max)}$ from BS 7671 = 7.67 ohms

$$Z_{test (max)} = \dfrac{1}{1.20} \times 7.67 = 6.39 \text{ ohms}$$

Therefore, as 0.55 is less than 6.39, the circuit is acceptable.

A

CITY OF LIVERPOOL COLLEGE
VAUXHALL ROAD
LIVERPOOL
L3 6BN

Appendix B
Resistance of copper and aluminium conductors

434.5.2
543.1.3
To check compliance with Regulation 434.5.2 and/or Regulation 543.1.3, i.e. to evaluate the equation $S^2 = I^2\,t/k^2$, it is necessary to establish the impedances of the circuit conductors to determine the fault current I and hence the protective device disconnection time t.

Fault current $I = U_0/Z_s$

where:

U_0 is the nominal voltage to Earth,

Z_s is the earth fault loop impedance, and

$Z_s = Z_e + R_1 + R_2$

where:

Z_e is that part of the earth fault loop impedance external to the circuit concerned

R_1 is the resistance of the line conductor from the origin of the circuit to the point of utilisation

R_2 is the resistance of the protective conductor from the origin of the circuit to the point of utilisation.

Similarly, in order to design circuits for compliance with the limiting values of earth fault loop impedance given in Tables 41.2 to 41.4 of BS 7671, it is necessary to establish the relevant impedances of the circuit conductors concerned at their operating temperature.

Table B1 gives values of $(R_1 + R_2)$ per metre for various combinations of conductors up to and including 50 mm^2 cross-sectional area. It also gives values of resistance (milliohms) per metre for each size of conductor. These values are at 20 °C.

CITY OF LIVERPOOL COLLEGE
VAUXHALL ROAD
LIVERPOOL
L3 6BN

▼ **Table B1** Values of resistance/metre for copper and aluminium conductors and of $(R_1 + R_2)$ per metre at 20 °C in milliohms/metre

Cross-sectional area (mm²)		Resistance/metre or $(R_1 + R_2)$/metre (mΩ/m)	
Line conductor	Protective conductor	Copper	Aluminium
1	–	18.10	
1	1	36.20	
1.5	–	12.10	
1.5	1	30.20	
1.5	1.5	24.20	
2.5	–	7.41	
2.5	1	25.51	
2.5	1.5	19.51	
2.5	2.5	14.82	
4	–	4.61	
4	1.5	16.71	
4	2.5	12.02	
4	4	9.22	
6	–	3.08	
6	2.5	10.49	
6	4	7.69	
6	6	6.16	
10	–	1.83	
10	4	6.44	
10	6	4.91	
10	10	3.66	
16	–	1.15	1.91
16	6	4.23	–
16	10	2.98	–
16	16	2.30	3.82
25	–	0.727	1.20
25	10	2.557	–
25	16	1.877	–
25	25	1.454	2.40
35	–	0.524	0.87
35	16	1.674	2.78
35	25	1.251	2.07
35	35	1.048	1.74
50	–	0.387	0.64
50	25	1.114	1.84
50	35	0.911	1.51
50	50	0.774	1.28

▼ Table B2 Ambient temperature multipliers (α) to Table B1

Expected ambient temperature (°C)	Correction factor*
0	0.92
5	0.94
10	0.96
15	0.98
20	1.00
30	1.04
40	1.08

* The correction factor is given by: {1 + 0.004 (ambient temp − 20 °C)} where 0.004 is the simplified resistance coefficient per °C at 20 °C given by BS 6360 for copper and aluminium conductors.

For verification purposes the designer will need to give the values of the line and circuit protective conductor resistances at the ambient temperature expected during the tests. This may be different from the reference temperature of 20 °C used for Table B1. The correction factors in Table B2 may be applied to the Table B1 values to take account of the ambient temperature (for test purposes only).

B1   Standard overcurrent devices

Table 41.2
Table 41.3
Table 41.4

Table B3 gives the multipliers to be applied to the values given in Table B1 for the purpose of calculating the resistance at maximum operating temperature of the line conductors and/or circuit protective conductors in order to determine compliance with the earth fault loop impedance of Table 41.2, 41.3 or 41.4 of BS 7671.

Where it is known that the actual operating temperature under normal load conditions is less than the maximum permissible value for the type of cable insulation concerned (as given in the tables of current-carrying capacity) the multipliers given in Table B3 may be reduced accordingly.

▼ Table B3 Conductor temperature factor F for standard devices

Multipliers to be applied to Table B1 for devices in Tables 41.2, 41.3, 41.4

Conductor installation	Conductor insulation		
	70 °C thermoplastic (PVC)	85 °C thermosetting (note 4)	90 °C thermosetting (note 4)
Not incorporated in a cable and not bunched (notes 1, 3)	1.04	1.04	1.04
Incorporated in a cable or bunched (notes 2, 3)	1.20	1.26	1.28

Notes:

Table 54.2

1 See Table 54.2 of BS 7671. These factors apply when protective conductor is not incorporated or bunched with cables, or for bare protective conductors in contact with cable covering.

Table 54.3

2 See Table 54.3 of BS 7671. These factors apply when the protective conductor is a core in a cable or is bunched with cables.

3 The factors are given by F = 1 + 0.004 {conductor operating temperature − 20 °C} where 0.004 is the simplified resistance coefficient per °C at 20 °C given in BS 6360 for copper and aluminium conductors.

4 If cable loading is such that the maximum operating temperature is 70 °C, thermoplastic (70 °C) factors are appropriate.

B2   Steel-wire armour, steel conduit and steel trunking

Formulae for the calculation of the resistance and inductive reactance values of the steel-wire armour of cables and of steel conduit, ducting and trunking are published in Chapter 6 of Guidance Note 6.

GN6

Generally, it is accepted that there is approximately a 10 °C difference between the conductor temperature and the outer sheath temperature for a steel-wire armoured cable at full load.

Index

A

Access   2.6.2r
Accuracy of test instruments   4.2
Additional protection   2.6.2h; 2.7.18
Additions and Alterations   1.4; 3.2; Chap 5

B

Barriers
  fire   2.6.2g
  protection by, during erection   2.7.4; 2.7.10
Basic protection   2.6.2i
Breaking capacity of devices   Table 2.8; 2.7.16
British Standards referenced
  Degrees of protection
    BS EN 60529   2.6.3; 2.7.10
  Electrical instruments
    BS EN 61010 (BS 5458)   1.1; 4.1
  Emergency lighting
    BS 5266   Table 3.2 note 3
  Fire alarms
    BS 5839   Table 3.2 note 5
  Instrument accuracy
    BS EN 61557 (BS 891)   Chap 4
  Residual current devices
    BS 4293, 7288   2.7.18, Table 2.9
    BS EN 61008, 61009   Table 2.7; 2.7.18, Table 2.9
  Switchgear
    BS EN 60947   2.6.3

C

Calibration (checking of accuracy)   4.2
Certificate, Electrical Installation   2.2; 2.7.2; Chap 5
Client   1.3; 3.8.2; 3.8.4
Colour coding (identification of conductors)   2.6.2b; 3.9.1
Competence   1.2; 3.8.1
Condition Report, Electrical Installation   1.3.1; 3.11; Chap 5
Construction (Design and Management)
    Regulations 1994   Introduction; 1.5; 3.4
Continuity
  main bonding   2.7.4; 2.7.5; Table 3.4; 3.10.3
  protective conductors   2.7.5; 3.10.3; Fig 2.1a; Fig 2.1b
    'all-insulated' installations –
      test method 1   2.7.5; Fig 2.1a

CITY OF LIVERPOOL COLLEGE
VAUXHALL ROAD
LIVERPOOL
L3 6BN

test method 2	2.7.5, Fig 2.1b
metallic enclosures as protective conductors	
test procedure	2.7.5
ring final circuit conductors, test method	2.7.4; 2.7.6; Fig 2.2; Fig 2.3
supplementary bonding	2.7.5; Table 3.3
testers	4.3

D

Defects, omissions	2.7.2
Designer	Introduction
Diagrams, presence of	2.6.2m

E

Earth electrode resistance	2.7.4; 2.7.13; Table 2.7; Table 3.4; Fig 2.7
Earth electrode resistance testers	2.7.13; 4.6
Earth fault loop impedance	
maximum measured values	Appx A
temperature correction	2.7.14; Table A6
test method	2.7.14
testers	4.5
Electrical Installation Certificate	2.2; 2.7.2; Chap 5
Electrical separation	
insulation resistance tests	2.7.9
source of supply	
inspection	2.7.9
measurement of voltage	2.7.9
Electricity at Work Regulations 1989	Preface; 2.1; Table 3.2 note 2; 3.3; 3.5; 4.9
Electricity Safety, Quality and Continuity Regulations 2002	Table 3.2 note 1

F

Fault protection	2.6.2h;
Fire barriers, provision of	2.6.2g
Forms	Chap 5
Functional extra-low voltage	Table 2.2
Functional testing	2.7.4; 2.7.19; Table 3.4; 3.10.3

G

General requirements	Chap 1

H

Health and Safety at Work etc. Act 1974	Introduction; 2.3; 3.4
Health and Safety Executive Guidance Note GS 38	1.1; 4.1
Housing (Scotland) Act 2006	1.3.2

I

Identification by colour	2.6.2b
Information for inspector	2.3; 3.6
Initial inspection	
general procedure	2.6.1
inspection checklist	2.6.3
inspection items	2.6.2
schedule	Chap 5

scope · 2.4
Initial testing · 2.7
  results, recording of · 2.7.1
  test sequence · 2.7.4
Initial verification · Chap 2
  certificates · 2.2; Chap 5
  frequency of subsequent inspection · 2.5
  information for inspector · 2.3
  purpose of · 2.1
  relevant criteria · 2.1
  scope · 2.4
  typical forms · Chap 5
Inspection checklist · 2.6.3
Inspection Schedule · Chap 5
Inspector's competence · 1.2; 3.8.1
Instruments · 1.1; Chap 4
Insulation resistance · 2.7.7; 3.10.3b; Table 2.2
Insulation resistance testers · 4.4

L
Labelling and marking · 2.6.2p,s; 3.9.1
Landlord and Tenant Act 1985 · 1.3.2; Table 3.2 note 10

M
Medical locations · 2.7.21
Minor Works Certificate · 2.2; Chap 5
Model forms · 2.7.3; Chap 5
Motor circuits · 3.10.3a

N
Non-conducting location · 2.7.11
Non-standard colours · 2.6.2p
Notices · 2.6.2p; 3.9.1

O
Ohmmeters
  insulation resistance · 2.7.7; 4.4
  low-resistance · 2.7.5; 2.7.6; 4.3
Older installations, inspection of · 3.12
Operational manual · Introduction

P
PELV · 2.6.2h; Table 2.2; 2.7.8; Table 2.5
Periodic inspection
  general procedure · 3.8.3
  inspection checklist · 3.9.1
  purpose of · 3.1; 3.8.1
  report · 1.3.1; 3.11; 5.3
  sampling · 3.8.4; Table 3.3
  scope · 3.8.1
  visual inspection · 3.8.2
Periodic inspection and testing
  forms · Chap 5

CITY OF LIVERPOOL COLLEGE
VAUXHALL ROAD
LIVERPOOL
L3 6BN

frequency of	3.7; Table 3.2
general procedure	3.8.2; 3.8.3
general requirements	Chap 1; Table 3.1
information for inspector	3.6
inspector's competence	1.2; 3.8.1
necessity for	3.2
purpose of	3.1; 3.8.1
record keeping	1.5
requirement for	3.2; 3.8.1
routine checks	3.5; Table 3.1; Table 3.2
Periodic testing	3.10; Table 3.4
Phase sequence	
instruments	4.8
verification of	2.7.17
Polarity	2.7.12; 3.10.3c; Table 3.4
test method	2.7.12; Fig 2.6
Prospective fault current	2.7.16
Protection by	
barriers provided during erection	2.7.4; 2.7.10
enclosures provided during erection	2.7.4; 2.7.10

R

Record keeping	1.5
Reference tests	
continuity of protective conductors	2.7.4; 2.7.5; Fig 2.1a; Fig 2.1b; Table 3.4; 3.10.3a
metallic enclosures as protective conductors	2.7.5
test methods 1 and 2	2.7.5
continuity of ring final circuit conductors	2.7.6; Fig 2.2; Fig 2.3; Table 3.4
insulation	
resistance test	2.7.7; 3.10.3b; Table 2.2
resistance of floors and walls	2.7.11
sequence of	2.7.4
Relevant criteria	2.1
Rented accommodation	1.3.2
Report, Electrical Installation Condition	1.3.1; 3.11; Chap 5
Residual current devices (RCDs)	
for additional protection	2.6.2h
notice	2.6.2p
test method	2.7.18; Table 3.4
testers	4.7
Resistance of conductors	Appx B
Routine checks	3.5; Table 3.1; Table 3.2
Rule of thumb	
earth fault loop impedance	2.7.15
3-phase prospective fault current	2.7.16

S

Safety during testing	1.1; 2.7.15; 3.8.2
Sampling	3.8.4; Table 3.3
Schedule of inspections	1.3.1; 2.7.2; 5.1; 5.3
Schedule of test results	1.3.1; 2.7.2; 5.1; 5.3
SELV	2.6.2h; 2.7.8; Table 2.2; Table 2.4
Separation of circuits	2.6.2h; 2.7.4; 2.7.8; 2.7.9

Shock hazards	1.1
Specialised systems	2.6.2h
Specification	Introduction
Surveying, thermographic	4.9

T

Test instruments	1.1; Chap 4
Thermal effects	2.6.2g,n; 3.9.1
Thermographic equipment	4.9

U

Utilisation categories, switchgear	2.6.2j; Table 2.1

V

Verification, initial	Chap 2
Voltage drop, verification	2.7.20

IET Wiring Regulations and associated publications

The IET prepares regulations for the safety of electrical installations, the IET Wiring Regulations (BS 7671: Requirements for Electrical Installations), which are the standard for the UK and many other countries. The IET also offers guidance around BS 7671 in the form of the Guidance Notes series and the Electrician's Guides as well as running a technical helpline. The Wiring Regulations and guidance are now also available as e-books through Wiring Regulations Digital (see overleaf).

IET Members receive discounts across IET publications and e-book packages.

Requirements for Electrical Installations IET Wiring Regulations 17th Edition (BS 7671:2008 incorporating Amendment No. 1:2011)
Order book PWR1701B Paperback 2011
ISBN: 978-1-84919-269-9 **£80**

On-Site Guide BS 7671:2008(2011)
Order book PWGO171B Paperback 2011
ISBN: 978-1-84919-287-3 **£24**

IET Guidance Notes

The IET also publishes a series of Guidance Notes enlarging upon and amplifying the particular requirements of a part of the IET Wiring Regulations.

Guidance Note 1: Selection & Erection, 6th Edition
Order book PWG1171B Paperback 2012
ISBN: 978-1-84919-271-2 **£32**

Guidance Note 2: Isolation & Switching, 6th Edition
Order book PWG2171B Paperback 2012
ISBN: 978-1-84919-273-6 **£27**

Guidance Note 3: Inspection & Testing, 6th Edition
Order book PWG3171B Paperback 2012
ISBN: 978-1-84919-275-0 **£27**

Guidance Note 4: Protection Against Fire, 6th Edition
Order book PWG4171B Paperback 2012
ISBN: 978-1-84919-277-4 **£27**

Guidance Note 5: Protection Against Electric Shock, 6th Edition
Order book PWG5171B Paperback 2012
ISBN: 978-1-84919-279-8 **£27**

Guidance Note 6: Protection Against Overcurrent, 6th Edition
Order book PWG6171B Paperback 2012
ISBN: 978-1-84919-281-1 **£27**

Guidance Note 7: Special Locations, 4th Edition
Order book PWG7171B Paperback 2012
ISBN: 978-1-84919-283-5 **£27**

Guidance Note 8: Earthing & Bonding, 2nd Edition
Order book PWG8171B Paperback 2012
ISBN: 978-1-84919-285-9 **£27**

Electrician's Guides

Electrician's Guide to the Building Regulations Part P, 2nd Edition
Order book PWGP170B Paperback 2008
ISBN: 978-0-86341-862-4 **£22**

continues overleaf ▶

Electrical Installation Design Guide
Order book PWR05030 Paperback 2008
ISBN: 978-0-86341-550-0   **£22**
To be updated 2012

Electrician's Guide to Emergency Lighting
Order book PWR05020 Paperback 2009
ISBN: 978-0-86341-551-7   **£22**

**Electrician's Guide to Fire Detection and
Alarm Systems**
Order book PWR05130 Paperback 2010
ISBN: 978-1-84919-130-2   **£22**

Other guidance

**Commentary on IEE Wiring Regulations
(17th Edition, BS 7671:2008)**
Order book PWR08640 Hardback 2009
ISBN: 978-0-86341-966-9   **£65**

Electrical Maintenance, 2nd Edition
Order book PWR05100 Paperback 2006
ISBN: 978-0-86341-563-0   **£40**

**Code of Practice for In-service Inspection and
Testing of Electrical Equipment, 3rd Edition**
Order book PWR08630 Paperback 2007
ISBN: 978-0-86341-833-4   **£40**

**Electrical Craft Principles, Volume 1,
5th Edition**
Order book PBNS0330 Paperback 2009
ISBN: 978-0-86341-932-4   **£25**

**Electrical Craft Principles, Volume 2,
5th Edition**
Order book PBNS0340 Paperback 2009
ISBN: 978-0-86341-933-1   **£25**

**For more information and to buy the IET
Wiring Regulations and associated guidance,
visit www.theiet.org/electrical**

Wiring Regulations Digital

The IET Wiring Regulations and associated guidance
are now available in e-book format. You can search
and link within and between books, make your own
notes and print pages from the books. There are
two packages available:

Wiring Regulations Digital: Domestic
BS 7671:2008(2011)*, the On-Site Guide, Guidance
Notes 1 & 3 and the Electrician's Guide to the
Building Regulations.   **£175 + VAT**

Wiring Regulations Digital: Industrial
BS 7671:2008(2011)*, the On-Site Guide, Guidance
Notes 1-8, the Electrician's Guide to the Building
Regulations and the Electrical Installation Design
Guide.   **£350 + VAT**

**Find out more and order at
www.theiet.org/digital-regs**
* You will initially receive the 17th Edition version of each of
these e-books and any published Amendment No. 1 titles
to date. Your package will be updated for free to each new
Amendment No. 1 title upon publication.

Electrical excellence

Order Form

How to order

BY PHONE:
+44 (0)1438 767328
BY FAX:
+44 (0)1438 767375
BY POST:
The Institution of
Engineering
and Technology,
PO Box 96,
Stevenage
SG1 2SD, UK
ONLINE:
www.theiet.org/electrical

INFORMATION SECURITY
Please do not submit your form by
email. The IET takes the security of your
personal details and credit/debit card
information very seriously and will not
process email transactions.

Free postage within the UK. Outside
UK (Europe) add £5.00 for first title
and £2.00 for each additional book.
Rest of World add £7.50 for the first
book and £2.00 for each additional
book. Books will be sent via airmail.
Courier rates are available on request,
please call +44 (0) 1438 767328
or email sales@theiet.org for rates.

**Member Discounts: These cannot be
used in conjunction with any other IET
discount offers.

GUARANTEED RIGHT OF RETURN:
If at all unsatisfied, you may return
books in new condition within 30 days
for a full refund. Please include a copy
of the invoice.

DATA PROTECTION:
The information that you provide to the
IET will be used to ensure we provide
you with products and services that
best meet your needs. This may include
the promotion of specific IET products
and services by post and/or electronic
means. By providing us with your email
address and/or mobile telephone
number you agree that we may contact
you by electronic means. You can
change this preference at any time
by visiting www.theiet.org/my.

Details

Name:

Job Title:

Company/Institution:

Address:

Postcode: Country:

Tel: Fax:

Email:

Membership No (if Institution member):

Ordering information

Quantity	Book No.	Title/Author	Price (£)
		CITY OF LIVERPOOL COLLEGE	
		VAUXHALL ROAD	
		LIVERPOOL	
		L3 6BN Subtotal	
		Member discount **	
		+ Postage /Handling*	
		Total	

Payment methods

☐ By **cheque** made payable to The Institution of Engineering and Technology

☐ By **credit/debit card:** ☐☐☐☐ ☐☐☐☐ ☐☐☐☐ ☐☐☐☐

☐ Visa   ☐ Mastercard   ☐ American Express   ☐ Maestro Issue No: _____

Valid from: ☐☐ ☐☐   Expiry Date: ☐☐ ☐☐   Card Security Code: ☐☐☐☐
                                                        (3 or 4 digits on reverse of card)

Signature _____ Date_____

Cardholder Name:

Cardholder Address:

Town: Postcode:

Country:

Tel:

Email:

The Institution of Engineering and Technology is registered as a Charity in England & Wales (no 211014) and Scotland (no SC038698). Michael Faraday House, Six Hills Way, Stevenage, SG1 2AY

Dedicated website

Everything that you need from the IET is now in one place. Ensure that you are up-to-date with BS 7671 and find guidance by visiting our dedicated website for the electrical industry.

Catch up with the latest forum discussions, download the Wiring Regulations forms (as listed in BS 7671) and read articles from the IET's **free** Wiring Matters magazine.

The IET Wiring Regulations BS 7671:2008(2011) and all associated guidance publications can be bought directly from the site. You can even pre-order titles that have not yet been published at discounted prices.
www.theiet.org/electrical

Membership

Passionate about your career? Become an IET Member and benefit from a range of benefits from the industry experts. As co-publishers of the IET Wiring Regulations, we can assist you in demonstrating your technical professional competence and support you with all your training and career development needs.

The Institution of Engineering and Technology is the professional home for life for engineers and technicians. With over 150,000 members in 127 countries, the IET is the largest professional body of engineers in Europe.

Joining the IET and having access to tailored products and services will become invaluable for your career and can be your first step towards professional qualifications.

You can take advantage of...

- a 35% discount on BS 7671:2008(2011) the IET Wiring Regulations, associated guidance publications and Wiring Regulations Digital
- career support services to assist throughout your professional development
- dedicated training courses, seminars and events covering a wide range of subjects and skills
- an array of specialist online communities
- professional development events covering a wide range of topics from essential business skills to time management, health and safety, life skills and many more
- access to over 100 Local Networks around the world
- live IET.tv event footage
- instant on-line access to over 70,000 books, 3,000 periodicals and full-text collections of electronic articles through the Virtual library, wherever you are in the world.

Join online today: **www.theiet.org/join** or contact our membership and customer service centre on +44 (0)1438 765678.

CITY OF LIVERPOOL COLLEGE
VAUXHALL ROAD
LIVERPOOL
L3 6BN